International Conference on Electrical Engineering and Control Technologies (CEECT 2019)

AA007646

IOP Conference Series: Materials Science and Engineering
Volume 752

Singapore
5 - 7 December 2019

ISBN: 978-1-7138-2659-0
ISSN: 1757-8981

Printed from e-media with permission by:

Curran Associates, Inc.
57 Morehouse Lane
Red Hook, NY 12571

Some format issues inherent in the e-media version may also appear in this print version.

This work is licensed under a Creative Commons Attribution 3.0 International Licence.
Licence details: http://creativecommons.org/licenses/by/3.0/.

No changes have been made to the content of these proceedings. There may be changes to pagination and minor adjustments for aesthetics.

Printed with permission by Curran Associates, Inc. (2021)

For permission requests, please contact the Institute of Physics at the address below.

Institute of Physics
Dirac House, Temple Back
Bristol BS1 6BE UK

Phone: 44 1 17 929 7481
Fax: 44 1 17 920 0979

techtracking@iop.org

Additional copies of this publication are available from:

Curran Associates, Inc.
57 Morehouse Lane
Red Hook, NY 12571 USA
Phone: 845-758-0400
Fax: 845-758-2633
Email: curran@proceedings.com
Web: www.proceedings.com

TABLE OF CONTENTS

APPLICATION OF CONTROL TECHNIQUES

AI APPLICATIONS IN POWER SYSTEMS

POWER SYSTEM OPERATION

PREFACE

CEECT 2019 was held in Singapore during Dec.5-7, 2019. CEECT 2019 is co-organized by Asia Pacific Institute of Science and Engineering(APISE) and Hong Kong Society of Mechanical Engineers(HKSME). This conference aims to provide a platform for researchers and engineers to share their ideas, recent developments and successful practices in Electrical Engineering and Control Technologies. The participants of the conference were from almost every part of the world, with background of either academia or industry, even well-known enterprise. The success and prosperity of the conference is reflected high level of the papers received.

The proceedings are a compilation of the accepted papers and represent an interesting outcome of the conference. This book covers 3 chapters: 1. Application of Control Techniques; 2. AI Applications in Power Systems; 3. Power System Operation.

We would like to acknowledge all of those who supported CEECT 2019. Each individual and institutional help were very important for the success of this conference. Especially we would like to thank the organizing committee for their valuable advices in the organization and helpful peer review of the papers.

We sincerely hope that CEECT 2019 will be a forum for excellent discussions that will put forward new ideas and promote collaborative researches. We are sure that the proceedings will serve as an important research source of references and the knowledge, which will lead to not only scientific and engineering progress but also other new products and processes.

Tek-Tjing Lie, Auckland University of Technology, New Zealand
Conference Chair

Content from this work may be used under the terms of the Creative Commons Attribution 3.0 licence. Any further distribution of this work must maintain attribution to the author(s) and the title of the work, journal citation and DOI.
Published under licence by IOP Publishing Ltd

IOP Conf. Series: Materials Science and Engineering **752** (2020) 011002 doi:10.1088/1757-899X/752/1/011002

COMMITTEES

Conference Chair

Tek-Tjing Lie, Auckland University of Technology, New Zealand

Program Chair

Anurag Sharma, Newcastle University, UK

Scientific Committee

Qianchuan Zhao, Tsinghua University, China

Organizing Committee

Ling Li, Tongji University, China
Rongyan Yu, Chengdu Huanyu Shida Conference Service Ltd., China

Local Committee

Yan Xu, Nanyang Technological University, Singapore

International Technical Committee

Chenn-Jung Huang, National Dong Hwa University, Taiwan, ROC
Chun Yi Su, Concordia University, Canada
Chunhua LIU, City Unviersity of Hong Kong, Hong Kong
Farhad Shahnia, Murdoch University, Australia
Loulin Huang, AUT University, Australia
Lei Chen, Wuhan University, China
Min Wang, Hohai University, China
Marcel ISTRATE, Gheorghe Asachi Technical University of Iasi County, Romania
Mohammed Haque, University of South Australia, Australia
Mohamed Yahia Edries, Space Division National Authority for Remote Sensing and Space Science, Egypt
Mohamed Dahidah, Newcastle University, UK
Ramesh Rayudu, Victoria University of Wellington, New Zealand
Yun Yang, The University of Hong Kong, Hong Kong, China

Content from this work may be used under the terms of the Creative Commons Attribution 3.0 licence. Any further distribution of this work must maintain attribution to the author(s) and the title of the work, journal citation and DOI.
Published under licence by IOP Publishing Ltd

IOP Conf. Series: Materials Science and Engineering **752** (2020) 011003 doi:10.1088/1757-899X/752/1/011003

Peer review statement

All papers published in this volume of *IOP Conference Series: Materials Science and Engineering* have been peer reviewed through processes administered by the proceedings Editors. Reviews were conducted by expert referees to the professional and scientific standards expected of a proceedings journal published by IOP Publishing.

Content from this work may be used under the terms of the Creative Commons Attribution 3.0 licence. Any further distribution of this work must maintain attribution to the author(s) and the title of the work, journal citation and DOI.
Published under licence by IOP Publishing Ltd

IOP Conf. Series: Materials Science and Engineering **752** (2020) 012001 doi:10.1088/1757-899X/752/1/012001

Multilayer Formation Control in Constrained Space

Guangfu Ma, Dongyu Li*, Chuanjiang Li and Wei Zhang

Guanfu Ma, Dongyu Li, and Chuanjiang Li are with the Department of Control Science and Engineering, Harbin Institute of Technology, Harbin 150001, China.
Dongyu Li is also with the Department of Electrical and Computer Engineering, National University of Singapore, Singapore 117576, Singapore.
Wei Zhang is with Shanghai Institute of Satellite Engineering, Shanghai 201109, China.

E-mail: `lidy@hit.edu.cn`

Abstract. This paper addresses the multilayer formation control (MLFC) problem for multi-agent systems in constrained space. A layered distributed finite-time estimator (LDFE) is proposed to acquire the target states for agents in each layer. To avoid collisions with borders, obstacles, as well as the other agents in the constrained space, an artificial potential function is designed based on the Dirac delta function. Based on the LDFE and the proposed potential function, the MLFC algorithm is proposed for multiple Euler-Lagrange systems (MELSs). The semi-global uniform ultimate boundedness of closed-loop errors is guaranteed by Lyapunov stability theory, while the desired formation of each layer can be achieved without collisions occurring in the constrained space. Simulation results are given to show the effectiveness of the proposed approaches.

1. Introduction

Multi-agent system (MAS) control includes interesting topics, such as the consensus problem [1, 2, 3, 4], the tracking problem [5, 6, 7, 8, 9], the formation problem [10, 11, 12, 13, 14, 15], and the containment problem [16, 17].

Recently, the formation-containment problem is defined by combining formation and containment problems [18, 19, 20, 21, 22]. Using a two-layer framework [23], the formation-containment problem is considered with leading following layers. Further, to articulate a more general formation problem, the multilayer formation control problem is proposed and defined in [24], and this framework is more in line with practical missions with a large number of agents. The agents in a layer are required to receive information from the prior layers, interact within the current layer, and transmit information to the subsequent layers. The agents in the first layer need to achieve a specific formation and the agents in the subsequent layers are expected to reach proper positions based on the primary formation shape formed by the prior layers.

In this paper, we extend the authors' previous result [24] with the consideration of control in constrained space, since a large number of agents should avoid collisions among each other as well as among unexpected obstacles or borders due to the lack of global environmental information. First, an LDFE is designed to estimate the desired state information for agents in each layer. Then, a novel artificial potential function is proposed with the aid of the Dirac delta function. Applying the potential function and the LDFEs, we finally introduce a model-based MLFC law for MELSs and all the closed-loop errors have been proved semi-globally uniformly ultimately bounded (SGUUB).

Content from this work may be used under the terms of the Creative Commons Attribution 3.0 licence. Any further distribution of this work must maintain attribution to the author(s) and the title of the work, journal citation and DOI.
Published under licence by IOP Publishing Ltd

2. Layered Formation Problem Formulations

Suppose that the multilayer formation consists of l layers with N_n agents in the nth layer, $n = 1, 2, \ldots, l$. Let $m = N_1 + N_2 + \ldots + N_l$. The agents are ordered by layers. Let V_{Li} denote the agent set of the nth layer as

$$V_{Li} = \left\{ \sum_{j=1}^{i-1} N_j + 1, \sum_{j=1}^{i-1} N_j + 2, \ldots, \sum_{j=1}^{i} N_j \right\}, \tag{1}$$

with $i = 1, 2, \ldots, l$, and Define $V_{LM} = V_{L1} \cup V_{L2} \cup \ldots \cup V_{Ll}$.

We assume the internal information flow in the nth layer, $n = 2, 3, \ldots, l$, is bidirectional, while the other information flow are all directed. Define layered Laplacian matrix $\mathcal{L} \in \mathbb{R}^{p(m+1)\times p(m+1)}$,

$$\mathcal{L} = \begin{bmatrix} 0 \otimes I_p & 0_{1\times N_1} \otimes I_p & 0_{1\times N_2} \otimes I_p & \cdots & \cdots & 0_{1\times N_l} \otimes I_p \\ \mathcal{L}_{L_1 2} \otimes I_p & \mathcal{L}_{L_1 1} \otimes I_p & 0_{N_1\times N_2} \otimes I_p & \cdots & \cdots & 0_{N_2\times N_l} \otimes I_p \\ 0_{N_2\times 1} \otimes I_p & \mathcal{B}_{L_2 2} & \mathcal{B}_{L_2 1} & \cdots & \cdots & 0_{N_3\times N_l} \otimes I_p \\ \vdots & \vdots & \vdots & \ddots & & \vdots \\ \vdots & \vdots & \vdots & & \mathcal{B}_{L_{l-1} 1} & 0_{N_{l-1}\times N_l} \otimes I_p \\ 0_{N_l\times 1} \otimes I_p & \mathcal{B}_{L_l l} & \mathcal{B}_{L_l l-1} & \cdots & \mathcal{B}_{L_l 2} & \mathcal{B}_{L_l 1} \end{bmatrix}, \tag{2}$$

where $\mathcal{L}_{L_1 1}$ represents the topology among the agents in the first layer, $\mathcal{B}_{F_n 1}$ the topology among the agents in the nth layer, $\mathcal{B}_{F_n 2}$ the topology among the agents in the $n-1^{\text{th}}$ layer and the agents in the nth layer.

Define a virtual agent 0, whose position and velocity are equal to the desired formation position and velocity. Then, virtual edges a_{i0}, $i = 1, 2, \ldots, N_1$, are given to denote the information flow between the ith agent in the first layer and agent 0.

Introduce an Laplacian matrix for the first layer and agent 0, where $\mathcal{L}_1 = [l_{ij}] \in \mathbb{R}^{N_1+1\times N_1+1}$, $l_{ii} = \sum_{j=1, j\neq i}^{N_1+1} a_{ij}$ and $l_{ij} = -a_{ij}$, $i \neq j$. A bearing-based approach is then given to describe the interaction in the nth layer, $n = 2, 3, \ldots, l$, and the bearing of agent j relative to agent i is given as [25]

$$g_{ij} = \frac{q_i - q_j}{\|q_i - q_j\|}, \tag{3}$$

$$P_{g_{ij}} = I_p - g_{ij} g_{ij}^T, \; i \in V_{Ln}, \; j \in V_{LM}, \tag{4}$$

where the desired inter-bearing is shown as g_{ij}^*. The Laplacian matrix $\mathcal{B}_{L_n v}$, $n = 2, 3, \ldots, l$, $v = 2, 3, \ldots, n$, is defined with the ijth block as

$$[\mathcal{B}_{L_n v}]_{ij} = \begin{cases} -P_{g_{ij}^*}, & i \neq j, \; i \in V_{Ln}, \; j \in V_{LM}, \\ \sum_{k\in\mathcal{N}_n} P_{g_{ik}^*}, & i = j, \; i \in V_{Ln}, \end{cases} \tag{5}$$

where $P_{g_{ij}^*} = I_d - g_{ij}^* g_{ij}^{*T}$.

The ith agent is described by the following equation

$$M_i(q_i)\ddot{q}_i + C_i(q_i, \dot{q}_i)\dot{q}_i + g_i(q_i) = \tau_i(t), \tag{6}$$

where $q_i \in \mathbb{R}^p$, $i = 1, 2, \ldots, m$ denotes the position vector of agent i, $M_i(q_i) \in \mathbb{R}^{p\times p}$ is the symmetric positive definite inertia matrix, $C_i(q_i, \dot{q}_i) \in \mathbb{R}^{p\times p}$ is the vector of Coriolis and centrifugal matrix, $g_i(q_i) \in \mathbb{R}^p$ denotes the gravitational force.

3. Layered Distributed Finite-Time Estimator

In this section, an LDFE is designed for agents in each layer. We propose the following LDFE protocol

$$\dot{\hat{q}}_i = \begin{cases} \hat{v}_i - \beta_1 \mathrm{sgn}\left[\sum_{j\in V_{L1}} a_{ij}\left(\hat{q}_i - \hat{q}_j\right) + a_{i0}\left(\hat{q}_i - q_0\right)\right], & i \in V_{L1}, \quad (7a)\\ \hat{v}_i = \vartheta_i + \dot{r}_i, & i \in V_{L1}, \quad (7b)\\ \dot{\vartheta}_i = -\beta_2 \mathrm{sgn}\left[\sum_{j\in V_{L1}} a_{ij}\left(\vartheta_i - \vartheta_j\right) + a_{i0}\left(\vartheta_i - \dot{q}_0\right)\right], & i \in V_{L1}, \quad (7c) \end{cases}$$

$$\dot{\hat{q}}_i = \begin{cases} \hat{v}_i - \beta_{2n-1} \mathrm{sgn}\left[\sum_{j\in V_{Ln}} P_{g^*_{ij}}\left(\hat{q}_i - \hat{q}_j\right) + \sum_{j\in V_{L(n-1)}\cup\ldots\cup V_{L1}} P_{g^*_{ij}}\left(\hat{q}_i - q_j\right)\right], & i \in V_{Ln}, \quad (8a)\\ \dot{\hat{v}}_i = -\beta_{2n} \mathrm{sgn}\left[\sum_{j\in V_{Ln}} P_{g^*_{ij}}\left(\hat{v}_i - \hat{v}_j\right) + \sum_{j\in V_{L(n-1)}\cup\ldots\cup V_{L1}} P_{g^*_{ij}}\left(\hat{v}_i - \dot{q}_j\right)\right], & i \in V_{Ln}, \quad (8b) \end{cases}$$

where $\beta_1, \beta_2, \ldots, \beta_{2n}$, $n = 2, 3, \ldots, l$ are positive constants. ϑ_i is an auxiliary variable. $\hat{q}_i$ and $\hat{v}_i$, $i \in V_{L1}$ are the estimates of the desired formation position and velocity for each agent in the first layer, respectively. Correspondingly, $\hat{q}_i$ and $\hat{v}_i$, $i \in V_{Ln}$ are the estimates of bearing-based positions and velocities of the desired formation for each agent in the nth layer, respectively. $r_i(t)$ is the auxiliary function decomposed from the desired formation configuration. Let q_0 and $\dot{q}_0$ denote the virtual agent 0's position and the velocity, respectively.

The following necessary assumptions are made.

Assumption 1 *Assume $N_1 \geq 2$, and there exists a directed spanning tree in the first layer with root agent 0.*

Assumption 2 *The acceleration of agent 0 is bounded by a constant δ_{L_1} as $\|\ddot{q}_0\| \leq \delta_{L_1} < \infty$.*

Assumption 3 *The target formation for the nth layer ($n = 2, 3, \ldots, l$) is infinitesimal bearing rigid, respectively.*

Define

$$\begin{aligned} Q_{L_1} &= \left[q_1^T, q_2^T, \ldots, q_{N_1}^T\right]^T, \\ C_{L_2} &= \left[c_{N_1+1}^T, c_{N_1+2}^T, \ldots, c_{N_1+N_2}^T\right]^T = -\left(L_{L_11}^{-1} L_{L_12} \otimes I_p\right) Q_{L_1}, \\ &\vdots \\ Q_{L_{n-1}} &= \left[q_1^T, q_2^T, \ldots, q_{N_1+\ldots+N_{n-2}+N_{n-1}}^T\right]^T, \\ C_{L_n} &= \left[c_{N_1+\ldots+N_{n-1}+1}^T, c_{N_1+\ldots+N_{n-1}+2}^T, \ldots, c_{N_1+\ldots+N_{n-1}+N_n}^T\right]^T \\ &= -\left(\mathcal{B}_{L_n1}^{-1}\left[\mathcal{B}_{L_nl}, \mathcal{B}_{L_nl-1}, \ldots, \mathcal{B}_{L_n2}\right]\right) Q_{L_{n-1}}, \end{aligned}$$

where $n = 2, 3, \ldots, l$,. $L_{L_11}, \mathcal{B}_{L_21}, \ldots, \mathcal{B}_{L_n1}$ are invertible based on Assumptions 1 and 3 [25][26].

Lemma 1 *[24] Let $\beta_1 > 0$ and $\beta_2 > \delta_{L_1}$. If Assumptions 1 and 2 hold, under the LDFEs* (7a) *and* (7b)*, the agents in the first layer can acquire the precise estimations of the desired formation velocity and position in finite time, respectively, which means*

$$\hat{v}_i(t) = \dot{q}_0(t) + \dot{r}_i(t),\ t \geq T_1, \quad (9)$$

$$\hat{q}_i(t) = q_0(t) + r_i(t),\ t \geq T_2,\ i \in V_{L1}, \quad (10)$$

where T_1 and T_2 are positive constants.

Therefore, $\hat{q}_i$ and $\hat{v}_i$, $i \in V_{L1}$ can be used to replace the position and velocity of the desired formation when $t \geq T_2$.

Next, consider the nth layer with $n = 2, 3, \ldots, l$.

Lemma 2 *[24] Let $\beta_{2n-1} > 0$ and $\beta_{2n} > \frac{\delta_{Ln} \| B_{Ln1}^{-1} B_{LnS} \|}{\lambda_{\min}(B_{Ln1})}$. If Assumption 3 holds, under the DFSEs* (8a) *and* (8b)*, the agents can acquire the precise estimations of the bearing-based velocity and position of the target formation in finite time, respectively, which means*

$$\hat{v}_i(t) = \dot{c}_i(t), \ t \geq T_{2n-1}, \tag{11}$$

$$\hat{q}_i(t) = c_i(t), \ t \geq T_{2n}, \ i \in V_{Ln}, \tag{12}$$

where T_{2n-1} and T_{2n} are positive constants.

Then, we can use $\hat{q}_i$ and $\hat{v}_i$, $i \in V_{Ln}$ to substitute the bearing-based weighted average of the positions and velocities of the agents in the $n - 1^{\text{th}}$ layer when $t \geq T_{2n}$.

Combining Lemma 1 and Lemma 2, we know that the agents in all layers can obtain desired position and velocity estimation in settling time $T_M = \max\{T_2, T_4, \ldots, T_{2n}\}$.

4. Control Design in Constrained Space

A critical issue for MAS control is to avoid the collision among agents and obstacles. Hence, real-time environment modeling techniques are essential to the agents, and they have to avoid collisions with pop-up threats without global environmental information.

To achieve collision avoidance among the agents, we design the potential functions for agent i as

$$P_{i,c}(\tilde{q}_{i,j}) = \sum_{j \in H_i} c_{i,j} \ln \frac{r - d_{safe}}{\|\tilde{q}_{i,j}\| - d_{safe}}, \tag{13}$$

where $\tilde{q}_{i,j} = q_i - q_j$, $(i, j \in V_{LM}, \ i \neq j)$, $c_{i,j} = c_{j,i}$ are positive constants, d_{safe} is the required safe distance, r is the danger detection range of each agent, and H_i represents all agents within the danger range of agent i as $H_i = \{j \in V_{LM}, j \neq i | \ \|\tilde{q}_{i,j}\| < r\}$.

To avoid collisions with obstacles in the constrained space, we define the constrained space as a point set $\Im = \Im_1 + \Im_2 + \ldots + \Im_R$, where R is the number of continuous regions. Define L_i as the edge of region $\Im_i$, $i = 1, 2, \ldots, R$. The irregular edges of spatial constraints are considered, where we divide an irregular edge into a number of segments. We assume that the edges of the constrained space are smooth. Using Dirac delta function [27, 28] to gather the potential forces from all segments of each edge in the constrained space, we construct the following potential function

$$P_{i,o}(q_i) = \sum_{x=1}^{R} \underbrace{\int_{q_i^{(1)}-r}^{q_i^{(1)}+r} \cdots \int_{q_i^{(p)}-r}^{q_i^{(p)}+r}}_{p} k_{cx} \frac{\delta(\|q_i - s - kP_x(s)\|)}{\|q_i - s\| - d_{safe}} \underbrace{\mathrm{d}s_1 \cdots \mathrm{d}s_p}_{p}, \tag{14}$$

where k is a real number, k_{cx} is a positive constant, $P_x(s)$ represents the perpendicular direction of L_x, and s_i denotes the ith element of vector s, $i \in 1, 2, \ldots, p$.

Next, define the following auxiliary variables,

$$z_{1i} = q_i - \hat{q}_i, \tag{15}$$

$$z_{2i} = \dot{q}_i - \alpha_{1i}, \ i \in V_{LM}, \tag{16}$$

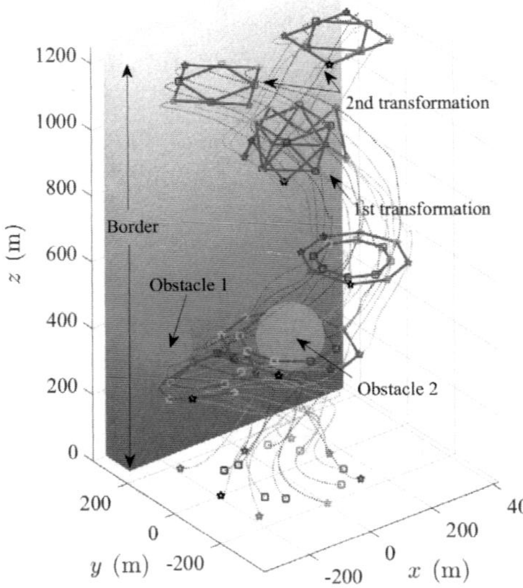

Figure 1: Trajectories of 16 satellites in the constrained space (3D view).

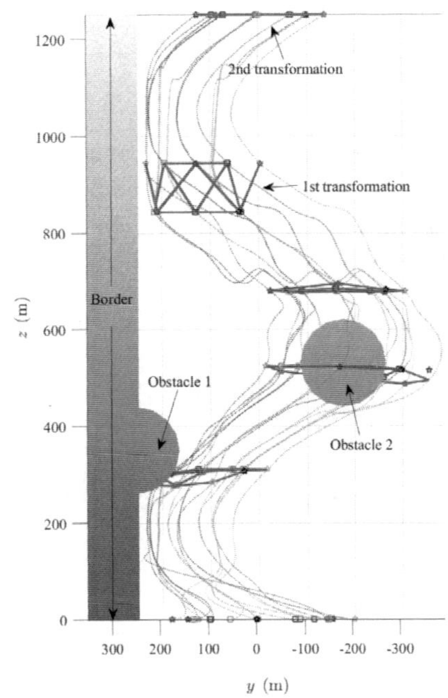

Figure 2: Trajectories of 16 satellites in the constrained space (side view).

where α_{1i} is the virtual control to be defined and $K_{1i} = K_{1i}^T > 0$ is the gain matrix.

We design the virtual control as

$$\alpha_{1i} = \hat{v}_i - (\Theta_i)^\dagger \left[K_{2i} V_{\chi i} + \left(\frac{\partial P_{i,o}}{\partial q_i} \right)^T \hat{v}_i + \left(\frac{\partial P_{i,c}}{\partial q_i} \right)^T (\hat{v}_i - \dot{q}_j) \right], \tag{17}$$

where $\Theta_i = z_{1i}^T + (\partial P_{i,o}/\partial q_i)^T + (\partial P_{i,c}/\partial q_i)^T$, $V_{\chi i} = \frac{1}{2} z_{1i}^T z_{1i} + P_{i,o} + P_{i,c}$, $K_{2i} = K_{2i}^T > 0$, and $(\Theta_i)^\dagger$ is the Moore-Penrose inverse of Θ_i defined as

$$\Theta_i(\Theta_i)^\dagger = \begin{cases} 0, & \Theta_i = [0, 0, \ldots, 0]^T, \\ 1, & \text{otherwise.} \end{cases} \tag{18}$$

Propose the following MLFC algorithm for each formation layer

$$\tau_i = -z_{1i} - K_{1i}\hat{z}_{2i} + M_i(q_i)\dot{\alpha}_{1i} + C_i(q_i, \dot{q}_i)\alpha_{1i} + g_i(q_i) - \frac{\partial P_{i,o}}{\partial q_i} - \frac{\partial P_{i,c}}{\partial q_i}, \tag{19}$$

where K_{1i} is the control gain matrix. Define the global closed-loop error signals as $z_1 = [z_{11}^T, z_{12}^T, \ldots, z_{1m}^T]^T$ and $z_2 = [z_{21}^T, z_{22}^T, \ldots, z_{2m}^T]^T$.

Theorem 1 *In each layer, under the MLFC algorithm* (19)*, the multilayer formation can be achieved within bounded errors without any collision, and the closed-loop errors* z_1 *and* z_2 *in each formation layer are SGUUB.*

Proof 1 *The proof can be given following the process in [11].*

5. Simulation

To show the performance of proposed MLFC algorithms, consider a group of 16 networked satellites with 8 satellites labeled as $1, \ldots, 8$ in the first layer, 4 satellites labeled as $9, \ldots, 12$ in the second layer, and 4 satellites labeled as $13, \ldots, 16$ in the third layer. The parameters of the satellites and the orbits are the same as those in [23]. The other parameters for simulation are omitted due to the limited pages.

The trajectories of satellites are given in Figures 1 and 2. In the constrained space, as shown in Figures 1 and 2, the collisions with the border, the obstacles, as well as the agents, are avoided successfully.

6. Conclusion

This paper using a multilayer framework to describe and extend the leader-follower problem with an arbitrary number of layers. A model-based MLFC algorithm is proposed with the consideration of collision avoidance. Under the MLFC law, the layered agents can achieve flexible formation configurations and transform the formation shape smoothly in the constrained space.

References

[1] Jadbabaie A, Lin J and Morse A S 2003 *IEEE Transactions on Automatic Control* **48** 988–1001 ISSN 0018-9286
[2] Olfati-Saber R and Murray R M 2004 *IEEE Transactions on Automatic Control* **49** 1520–1533 ISSN 0018-9286
[3] Ren W and Beard R W 2005 *IEEE Transactions on Automatic Control* **50** 655–661
[4] Yaghmaie F A, Su R, Lewis F L and Xie L 2017 *IEEE Transactions on Automatic Control* **62** 5578–5589
[5] Cao Y and Ren W 2012 *IEEE Transactions on Automatic Control* **57** 33–48
[6] Peng K and Yang Y 2009 *Physica A Statistical Mechanics & Its Applications* **388** 193–208
[7] Li D, Ma G, Li C, He W, Mei J and Ge S S 2018 *IEEE Transactions on Aerospace and Electronic Systems* **54** 2233–2245
[8] Hong Y, Chen G and Bushnell L 2008 *Automatica* **44** 846–850
[9] Li Z, Liu X, Ren W and Xie L 2013 *IEEE Transactions on Automatic Control* **58** 518–523
[10] Do K D 2007 *IEEE Transactions on Automatic Control* **52** 569–576
[11] Li D, Ge S S, Ma G and He W 2020 *International Journal of Control, Automation and Systems, in press, DOI: 10.1007/s12555-019-0380-2*
[12] Cheah C C, Hou S P and Slotine J J E 2009 *Automatica* **45** 2406–2411
[13] Li D, Ma G, He W, Ge S S and Lee T H *IEEE Transactions on Cybernetics, in press, DOI: 10.1109/TCYB.2019.2923119*
[14] Han Z, Guo K, Xie L and Lin Z 2018 *IEEE Transactions on Automatic Control* **64** 20–34
[15] Li X and Xie L 2018 *IEEE Transactions on Automatic Control* **9286** 1–14 ISSN 00189286
[16] Notarstefano G, Egerstedt M and Haque M 2011 *Automatica* **47** 1035–1040
[17] Ji M, Ferrari-Trecate G, Egerstedt M and Buffa A 2008 *IEEE Transactions on Automatic Control* **53** 1972–1975
[18] Dong X, Li Q, Ren Z and Zhong Y 2016 *Journal of the Franklin Institute* **353** 322–344 ISSN 00160032
[19] Dong X, Shi Z, Lu G and Zhong Y 2015 *International Journal of Robust and Nonlinear Control* **25** 3439–3456
[20] Wang Y, Liu X and Xiao J 2017 *Journal of the Franklin Institute* **354** 392–414 ISSN 0016-0032
[21] Wang Y, Song Y and Krstic M 2017 *IEEE transactions on cybernetics* **47** 2161–2172
[22] Wang Y W, Liu X K, Xiao J W and Shen Y 2018 *Automatica* **93** 26–32
[23] Li D, Zhang W, He W, Li C and Ge S S 2019 *IEEE Transactions on Cybernetics* **49** 675–687 ISSN 2168-2267
[24] Li D, Ge S S, He W, Ma G and Xie L 2019 *Automatica* **109** 108558 ISSN 0005-1098
[25] Zhao S and Zelazo D 2017 *IEEE Transactions on Control of Network Systems* **4** 429–438 ISSN 23255870 (*Preprint* `1506.05636`)
[26] Mei J, Ren W and Ma G 2012 *Automatica* **48** 653–659 ISSN 00051098
[27] Dirac P A M 1981 *The principles of quantum mechanics* 27 (Oxford university press)
[28] Ge S S, Liu X, Goh C H and Xu L 2016 *IEEE Transactions on Control Systems Technology* **24** 992–1003

IOP Conf. Series: Materials Science and Engineering **752** (2020) 012002 doi:10.1088/1757-899X/752/1/012002

Tracking Error Fitness Function V/F Control of Micro-Grid Based on Genetic Algorithm

Leiming Ma, Lingfei Xiao* and Min Xu

College of Energy and Power Engineering, Nanjing University of Aeronautics and Astronautics, China

*lfxiao@nuaa.edu.cn

Abstract. This paper presents a method of V/f control parameters tuning for micro-grid based on genetic algorithm. Firstly, the fitness function which representing the tracking error of V/f control system in micro-grid is established. The solving path of genetic algorithm is optimized by using the offset threshold condition between the output voltage and the instruction voltage of the load side, leading to that the individuals with high priority will tend to the optimal solution in the target solution, obtaining the global optimal V/f control parameters. Secondly, the frequency domain response characteristics of equivalent output impedance of V/f control system are analyzed, and the influence of PI parameters on the whole control system is compared. The results shows that the V/f control parameters tuning based on genetic algorithm with the tracking error fitness function can effectively guarantee a constant output of micro-grid's voltage and frequency. The effectiveness and the advantages of the method is verified by simulation result, which indicated that the proposed method is easy to implement and has a wide range of applications.

1. Introduction

Electricity, the pillar of national energy and economic lifeline, is particularly important in the rapid development of society. At present, power production mainly relies on non-renewable energy. Correspondingly, power distribution mainly relies on high-voltage long-distance transmission. However, the drawbacks of this mode appear as time goes by, which will not only waste a lot of electric energy, but also lead to resource exhaustion and environmental pollution [1]. By contrast, distributed generation has the advantages of in-situ power output, in-situ consumption, with no need for high-voltage transmission system, which can utilize clean secondary energy such as solar energy, wind energy and so on. Nevertheless, if the distributed power supply is connected to the power grid, it will bring a series of impacts on the distribution network, even impacting the public power grid and causing regional paralysis [2]. Therefore, it is of great significance to study the control method of micro-grid with distributed generation. At present, the control object of micro-grid mainly focuses on inverters, and the control strategies mainly include PQ method, V/f method and Droop method [3]. This paper mainly studies V/f method of micro-grid.

In reference [4], the control strategy of grid-connected inverters based on gain scheduling adaptive is proposed, and the simulation model and experimental platform of three-phase LCL grid-connected inverters are built to explain the feasibility and effectiveness of the adaptive control method. In reference [5], the conditions for maintaining the stable operation of micro-grid and the control strategies of PQ method and V/f method are studied when micro-grid is connected to the grid under

Content from this work may be used under the terms of the Creative Commons Attribution 3.0 licence. Any further distribution of this work must maintain attribution to the author(s) and the title of the work, journal citation and DOI.
Published under licence by IOP Publishing Ltd

various types of small signal disturbances. In reference [6], the V/f method based on droop characteristics is described comprehensively, and the constant voltage-frequency is achieved when the micro-grid operates on islands. In reference [7], under the effect of master-slave control of inverters, V/f method and PQ method are used to realize seamless switching of DGS, and synchronous phase-locked technology is introduced to prevent oscillation caused by sudden change of phase angle. In reference [8], the pole assignment method based on the transfer function of the control system is used to design the parameters of the micro-grid's PQ method and V/f method models respectively. Finally, the accuracy of the parameters is verified by simulation.

Whether PQ method, V/f method or Droop method are realized by PI regulation, so it is very important to set the parameters of PI regulator accurately. In classical control theory, the methods commonly used in double-loop control system in synchronous coordinate system are classical system tuning method and closed-loop pole placement method. The above methods are based on the transfer function of the double-loop method system. Although the stability of the control can be ensured, the influence of the closed-loop zeros on the response of the control system is neglected. Therefore, necessary adjustments should be made according to the actual dynamic characteristics of the system to meet the requirements of the control performance index [9]. The parameter tuning method of PI controller based on genetic algorithm proposed in this paper does not based on accurate transfer function. It can directly adjust the parameters and realize the stability of the control system.

2. Micro-grid structure

Compared with the general power grid, the structure of micro-grid is more flexible. The internal load of micro-grid is divided into different grades, and different number of distributed generators are equipped according to different load levels. When distributed power generation (DG) operates independently, the voltage and frequency of the whole micro-grid are required to be constant. When DG is connected to the grid, the frequency of the output power voltage should be controlled and maintained in accordance with the large power grid [10]. At present, the output of distributed generation is mostly direct current, which needs to be converted to 50Hz AC power supply or grid-connected through DC/AC inversion. Therefore, power electronics technology, especially inverter technology, plays an unique role in the operation and control of micro-grid [11]. Structure of the general micro-grid shown in figure 1.

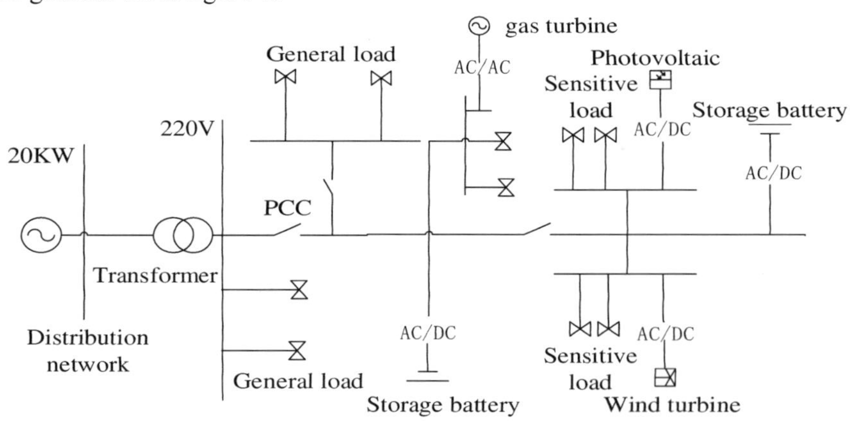

Figure 1. Structure of the typical micro-grid.

3. Design of V/f control method based on genetic algorithm for micro-grid

3.1. V/f control principle of micro-grid

Based on the voltage and current double-loop V/f control strategy, the control structure is shown in figure 2. In outer loop, the voltage control needs to collect the voltage U on load side. U_d and U_q are obtained via coordinate transformation and get the differences from the given values U_{dref} and U_{qref}. The reference values I_{dref} and I_{qref} of current inner loop are obtained by PI compensation. i_d and i_q are obtained by collecting the filter inductance current, which are inferior to I_{dref} and I_{qref} respectively by coordinate transformation. Through proportional compensation and feed-forward decoupling, the reference voltage u_d^* and u_q^* are obtained. Through coordinate inverse transformation and PWM modulation, the switching signal of the inverter can be achieved, ensuring the constant voltage amplitude and frequency [12].

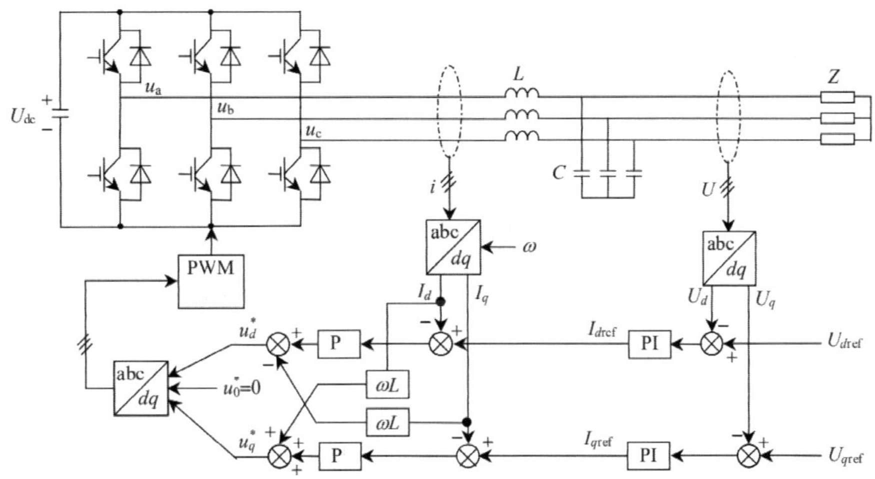

Figure 2. V/f control strategy.

Based on Kirchhoff's law, the voltage equation of filter inductance and the differential equation of filter capacitance can be obtained from figure 2.

$$L\frac{dI_L}{\mathrm{dt}} = U_{dc} - U_{ld} \tag{1}$$

$$C\frac{dU_{ld}}{dt} = I_L - \left(I_{ld} + I_f\right) \tag{2}$$

According to equation (1) and equation (2), a double-loop controller can be designed as shown in figure 3. Among them, the current loop is inner loop, the voltage loop is outer loop, and the current loop is proportional controller. The main purpose of the voltage loop is to maintain the load voltage stability, and the purpose of the current loop is to improve the dynamic response of the system.

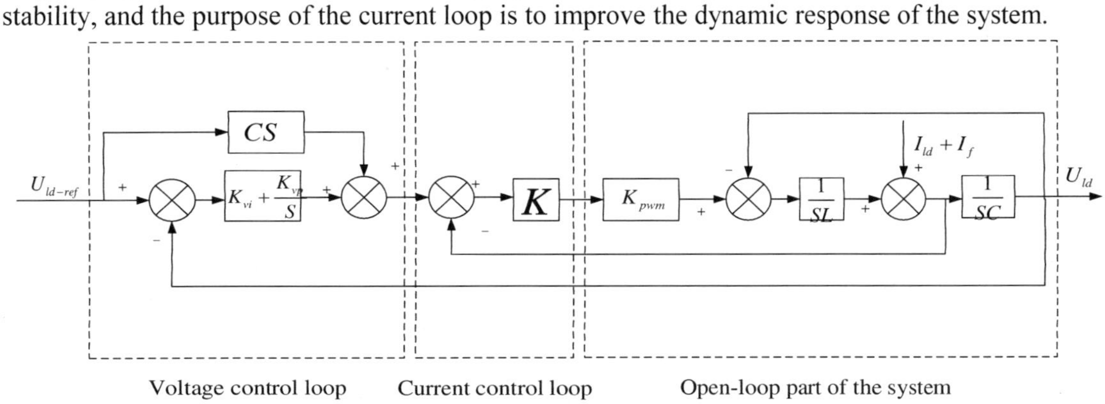

Figure 3. Double loop control of V/f.

The equivalent output impedance equation of the corresponding inverters are as follows:

$$Z(s)=\frac{Ls^2}{LCs^3+KK_{pwm}Cs^2+(1+KK_{pwm}K_{vp})s+KK_{pwm}K_{vi}} \quad (3)$$

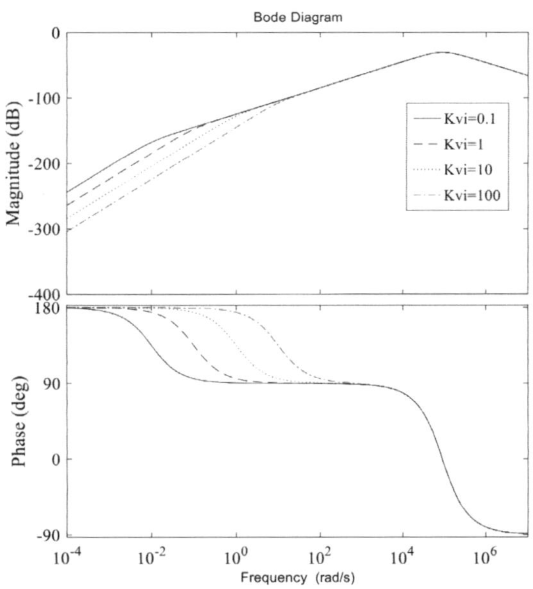

Figure 4. Frequency response, $K_{vp}=10$.

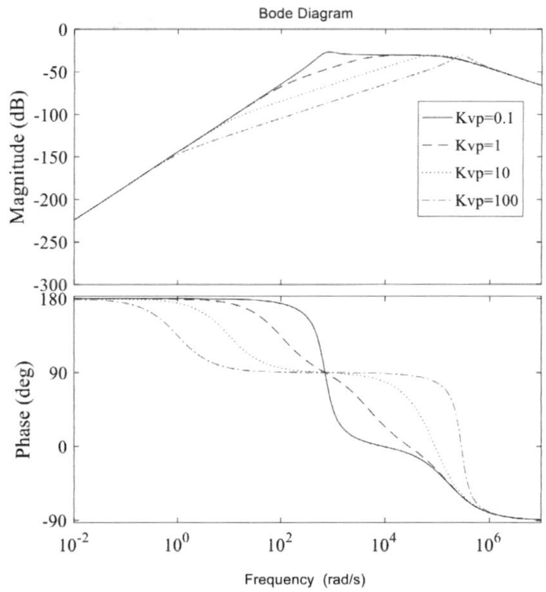

Figure 5. Frequency response, $K_{vi}=100$.

By changing K_{vp} and K_{vi}, the frequency response curve of equivalent output impedance was observed. When $K_{vp}=10$, the frequency response curves of K_{vi} at 0.1, 1, 10 and 100 are shown in figure 4. When $K_{vi}=100$, the frequency response curves of K_{vp} at 0.1, 1, 10 and 100 are shown in figure 5. It can be seen from figure 4 that the output impedance has broader inductance frequency band in consistent with the decreasing of K_{vi}.And it can be seen from figure 5 that the output impedance has broader inductance frequency band in consistent with the increasing of K_{vp}. Because of the fact that the high frequency output impedance resistivity will effectively suppress harmonics, the K_{vi} and K_{vp} parameter which have too wide inductive frequency band of output impedance can not be selected.

3.2. Design of genetic algorithm controller

(1).Design of fitness function

Generally, the solution corresponding to the individual in the population is directly taken as a component of the fitness function [13]. However, the genetic algorithm in this paper is not used to find the solution of a certain function, but to find the optimal parameters in V/f controller of micro-grid. Therefore, the selection of fitness function should satisfy the control requirement of micro-grid load side output voltage as close as possible to the command output voltage. The fitness function is designed as follows:

$$\min f=\frac{1}{2}e^2=\frac{1}{2}[y_1(t)-y(t)]^2 \quad (4)$$

In (4), $y(t)$ is three-phase symmetrical voltage $u=311\sin(100\pi)\,V$; $y_1(t)$ is load side output voltage; e represents tracking error, and the control effect is mainly indicated by the value of e.

(2).Parametric optimization principle of V/f controller based on genetic algorithm

The V/f control principle based on genetic algorithm is shown in figure 6.

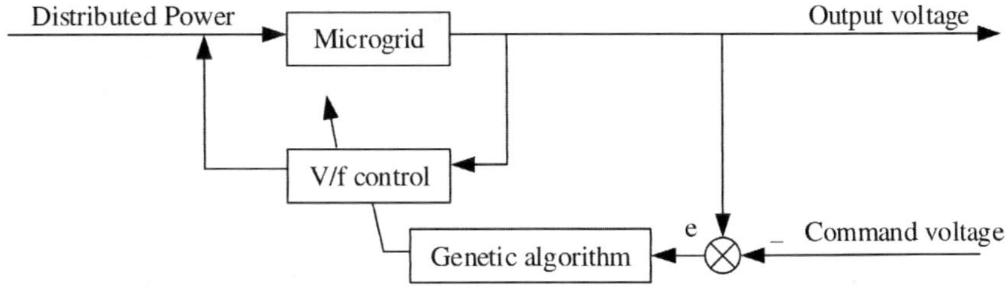

Figure 6. V/f control structure based on genetic algorithm.

According to figure 2, the V/f control simulation model of micro-grid is built, and the PI parameters of current loop are introduced into genetic algorithm for optimization. The parameters range K_p and K_i of PI regulator are chosen as [0,100]. The simulation time of each iteration is 1.5s. The operation parameters of the genetic algorithm are determined as follows: the initial population size is 300, the number of iterations is 9, the crossover probability is 0.4, and the mutation probability is 0.04. According to solution of genetic algorithm, the global optimal solution to ensure the stability of V/f controller is $K_p = 15.9574$ and $K_i = 98.291$. The results are consistent with the previous analysis of equivalent output impedance curve in frequency domain.

4. Simulation

Combining with V/f control structure figure 2, Simulation are carried out for islanding operation of micro-grid. The parameters of control system are set as follows: DC input $U_{dc} = 650V$, $r = 0.01\Omega$, filter capacitance and inductance $C = 200\mu F$, $L = 2mH$, $K = 1$, respectively. The total simulation time is set to 1.5s. Initially, the load power is $P = 100kW$, $Q = 0$, the load is cut into 60KW at 0.5s and the load is cut out 60KW at 1s. At this time, the inverters are controlled by V/f method. The voltage, current and frequency of the load side are shown in figure 7.

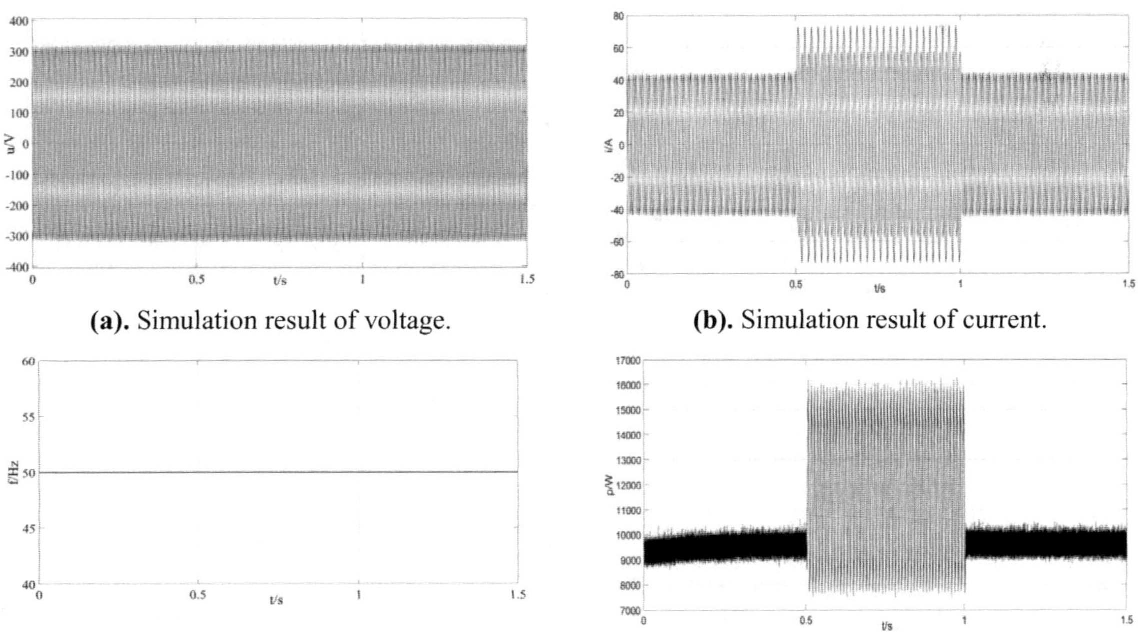

(a). Simulation result of voltage. **(b).** Simulation result of current.

(c). Simulation result of frequency. **(d).** Simulation result of power.

Figure 7. Simulation results of V/f control.

As shown in figure 7, the V/f control of the inverters can ensure that the phase voltage at the load side of the micro-grid is maintained at 311V and the frequency is maintained at 50Hz. At 0.5s, the load side current and frequency both have abrupt changes, which are due to the cut-in load. When the load is cut out at 1s, the load side current and power are back to the initial value. The above results prove the validity of V/f control and the feasibility of applying genetic algorithm to PI parameter adjustment.

5. Conclusion

In this paper, the V/f control of micro-grid under islanded operation mode is established, and the PI parameters of current loop are optimized by genetic algorithm. Based on the established fitness function representing tracking error, the PI parameters are solved by using the threshold condition between the output voltage and the instruction voltage of the load side. The global optimal PI parameters obtained by genetic algorithm are applied to the V/f control simulation model. The simulation results show that the method is effective. At the same time, the optimal value of V/f control parameters is also consistent with the analysis results of frequency domain response characteristics of equivalent output impedance, which can effectively suppress harmonics and improve control accuracy.

Acknowledgments

Funding: This research was funded by National Natural Science Foundation of China, grant number 51876089.

References

[1] Khan A A, Rehanim M H and Reisslein M 2016 Cognitive radio for smart grids: survey of architectures, spectrum sensing mechanisms, and networking protocols. *IEEE Communication Survey & Tutorials*, 18(1):860–98.
[2] Sen S, Kumar V 2018 Micro-grid control: a comprehensive survey. *Annual Reviews in control*, 45:118–51.
[3] Chen S, Song S, Li L and Shen J 2009 Survey on smart grid technology. *Power System Technology*, 33(8):1–7.
[4] Cespedes M, Jian S 2014 Adaptive control of grid-connected inverters based on online grid impedance measurements. *IEEE Transactions on Sustainable Energy*, 5(2):516-23.
[5] Venkataramanan G, IIlindala M 2007 Small signal dynamics of inverter interfaced distributed generation in a chain micro-grid. *IEEE Power Engineering Society General Meeting*, pp 1-6.
[6] Simpson-Porco J W, Simpson P, Dörfler F and Bullo F 2013 Synchronization and power sharing for droop-controlled inverters in islanded micro-grids. *Automatica*, 49:2603–11.
[7] Zhou H, Bhattacharya T, Tran D and Ashwin M 2011 Composite energy storage system involving battery and ultra-capacitor with dynamic energy management in micro-grid applications. *IEEE Transactions on Power Electronics*, 26(3):923–30.
[8] Zhang W, Liu L and Yang K 2014 Design of the power control system for micro-grid. *Process Automation Instrumentation*, 35(12):13–8.
[9] Pecas Lopes J A, Moreira C L and Madureira A G 2006 Defining control strategies for micro-grids islanded operation. *IEEE Transaction on Power systems*, 21(2):916-24.
[10] Zhou L, Huang Y, Guo K and Feng Y 2011 A survey of energy storage technology for micro-grid. *Power System protection and control*, 39(7):147–52.
[11] Hiroaki K, Yushi M and Toshifumi I 2010 Low-voltage bipolar-type DC micro-grid for super high quality distribution. *IEEE Transactions on power electronics*, 25(12):3066–75.
[12] Zhou X, Rong F, Lü Z, Luo A and Peng S 2012 A coordinate rotational transformation based virtual power V/f droop control method for low voltage micro-grid, *Automation of Electric Power Systems*, 36(2):47–51.
[13] Hamblin S 2013 On the practical usage of genetic algorithms in ecology and evolution. *Methods in Ecology and Evolution*, 4(2):184–94.

IOP Conf. Series: Materials Science and Engineering **752** (2020) 012003 doi:10.1088/1757-899X/752/1/012003

Power Stability Control Strategy of the WPT System Based on Load Detection

Ming Zhang*, **Xueliang Huang, Linlin Tan and Han Liu**

School of Electrical Engineering, Southeast University, Nanjing 210096, China

*zhangming1212@126.com

Abstract. The output power fluctuates when offset occurs between the receiving coil and the transmitting coil or load equivalent impedance changes in a wireless power transfer system (WPT). These two conditions will affect the safety of the system and the life of the battery load. Therefore, effective control methods are needed to ensure the stability of the load power. This paper analyzes the WPT system with LCC/S compensation topology and the secondary side containing DC/DC circuit. The relationship between mutual inductance and load voltage and current, mutual inductance and the system transmission efficiency, DC/DC circuit duty ratio and load voltage and current are deduced. The output power stability control strategy through load voltage and load current detection and control of DC/DC circuit is proposed. Finally, the WPT system with 1 kW output power is built through simulation and experiment, which verifies the effectiveness and correctness of theoretical analysis and control strategy in this paper.

1. Introduction

The WPT technology has been applied in implantable bioelectronics, portable electronic devices, unmanned aerial vehicles, electric vehicles and inspection robots in recent years [1-3]. The coupling coefficient and load impedance of WPT system are not fixed and may change during power supply, and these two parameters have a great influence on the output power. For example, when an electric vehicle drives to a wireless charge parking space, it cannot park in exactly the same place each time. There will be a certain deviation between the output power and the demand power if no measures are taken. This problem is more obvious in the application of dynamic wireless charging of electric vehicles. The output power fluctuates with the change of electric vehicle position in the process of dynamic wireless charging [4-5]. This situation may lead to the battery has been in charge and discharge switch state, which will have a greater impact on the battery life. In addition, the equivalent impedance of the battery load will also change during the charging process, which also can cause the output power fluctuation. Therefore, an effective output power stability control method needs to be proposed when there is deviation between the receiving coil and the transmitting coil or load equivalent impedance change.

There are many methods have been used to achieve the goal of stable output power. For example, the magnetic coupling mechanism is optimized to achieve more uniform distribution of magnetic field, the power source voltage is adjusted through primary and secondary real-time communication, and multiple transmitting coils are used for coordinated control to achieve stable output power [6-9]. An integrated control method of load estimation and power tracking is proposed in [10] to achieve constant power charging. In [11], a control strategy regarding two source voltages is investigated in

Content from this work may be used under the terms of the Creative Commons Attribution 3.0 licence. Any further distribution of this work must maintain attribution to the author(s) and the title of the work, journal citation and DOI.
Published under licence by IOP Publishing Ltd

order to minimize output power fluctuation. [12] presents a detailed deduction process to acquire the optimal coupling coefficient, tuned at which the system can achieve minimum output voltage fluctuation. The authors of [13] proposed the three-coil structure WPT system to achieve stable CC and CV charging output characteristics for battery charging applications.

The paper is arranged as follows. Section 2 analysis the WPT system based on the compensation topology of LCC/S and the secondary side containing the DC/DC circuit. Section 3 proposes a power stability control strategy for load current and voltage detection. In section 4, the control strategy is verified by simulation and experiment. Finally, section 5 draws conclusions from the research.

2. Circuit analysis

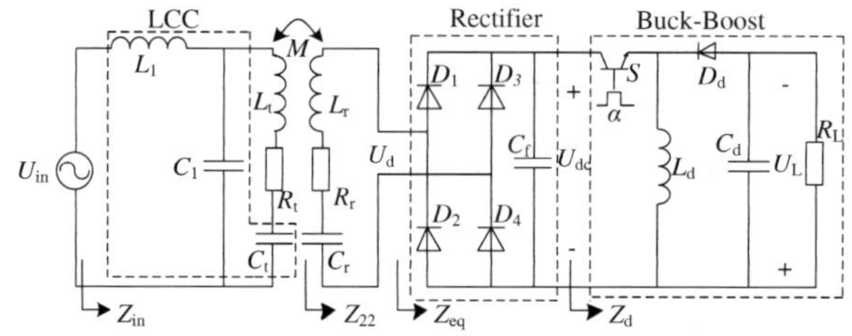

Figure 1. The WPT system with LCC/S compensation topology and secondary side containing DC/DC circuit.

The WPT system based on the compensation topology of LCC/S and the secondary side containing the DC/DC circuit is shown in figure 1. The compensation topology of the primary side is LCC structure, and the capacitor of the secondary side forms a series compensation structure with the receiving coil. After the compensation topology of the secondary side is the rectifying circuit, and then supplies power to the load through the cascade DC/DC circuit. In this paper, Buck-Boost circuit is adopted for DC/DC circuit, and the equivalent impedance Z_{eq} of rectifier filter circuit and Buck-Boost circuit is:

$$Z_{eq} = 8\pi^{-2} R_L (\alpha^{-1} - 1)^2 \tag{1}$$

where R_L is the load impedancec, α is the duty ratio of the Buck-Boost circuit. In figure 1, L_t and L_r are the inductance of transmitting coil and receiving coil respectively. R_t and R_r are the internal resistance of transmitting coil and receiving coil, respectively. C_t is connected in series with L_t, C_1 is in parallel with the series circuit of L_t and C_t, and L_1 is in series with this circuit to form the LCC compensation topology. C_r and L_r are connected in series to form S compensation topology.

To ensure that the input impedance of the system is pure impedance, the configuration of parameters should be satisfied the relation $\omega L_1 = \omega L_t - 1/\omega L_t = 1/\omega C_1$ (ω is the operating angular frequency). The secondary side should satisfy the relation $\omega L_r = 1/\omega C_r$ in the resonance state. Then the input impedance $\boldsymbol{Z}_{in}$ in figure 1 can be derived.

$$\boldsymbol{Z}_{in} = \omega^2 L_1^2 (Z_{ref} + R_t)^{-1} \tag{2}$$

From the input impedance, the input current $\boldsymbol{I}_{in}$ is

$$\boldsymbol{I}_{in} = \boldsymbol{U}_{in} (Z_{ref} + R_t)(\omega^2 L_1^2)^{-1} \tag{3}$$

Then, the current $\boldsymbol{I}_t$ of the transmitter coil can be deduced to be

$$\boldsymbol{I}_t = \boldsymbol{U}_{in} (j\omega L_1)^{-1} \tag{4}$$

The voltage $\boldsymbol{U}_d$ in figure 1 can be obtained from $\boldsymbol{I}_t$.

$$\boldsymbol{U}_d = M\boldsymbol{U}_{in} \left(L_1 (R_r Z_{eq}^{-1} + 1) \right)^{-1} \tag{5}$$

Ignoring the internal resistance of the power source, the system transmission efficiency formula can be obtained as follows:

$$\eta = \left(\left(R_t R_r^2 (\omega^2 M^2)^{-1} + R_r \right) Z_{eq}^{-1} + R_t Z_{eq} (\omega^2 M^2)^{-1} + 2 R_t R_r (\omega^2 M^2)^{-1} + 1 \right)^{-1} \tag{6}$$

On the contrary, the expression of mutual inductance can be obtained from equation (6).

$$M = \sqrt{R_r^2 R_t + Z_{eq}^2 R_t + 2 R_t R_r Z_{eq} \left((Z_{eq}(\eta^{-1} - 1) - R_r)\omega^2 \right)^{-1}} \tag{7}$$

The relation curves of load equivalent resistance and system efficiency, mutual inductance and efficiency is drawn by equation (6), as shown in figure 2 and figure 3. The transmission efficiency increases first and then decreases with the increase of equivalent load. The greater the mutual inductance, the higher the transmission efficiency. And the higher the equivalent impedance is, the faster the transmission efficiency increases with mutual inductance.

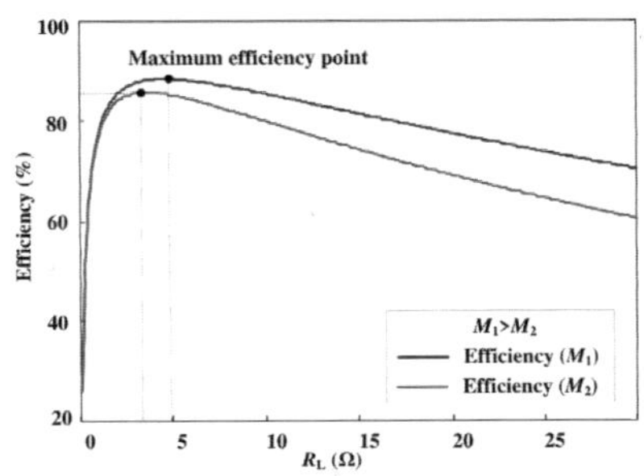

Figure 2. Curve of efficiency and load impedance.

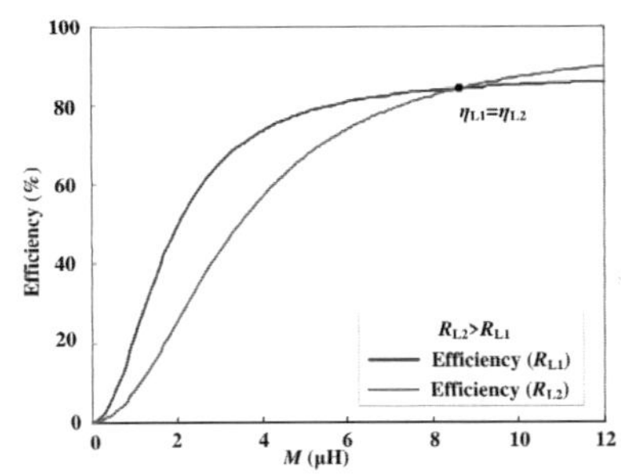

Figure 3. Curve of efficiency and mutual inductance.

When equivalent impedance meets $Z_{eq} = \sqrt{R_r^2 + \omega^2 M^2 R_r R_t^{-1}}$ and mutual inductance is constant, the maximum value of transmission efficiency η_m can be calculated.

$$\eta_m = \frac{\omega^2 M^2}{2R_t(\sqrt{R_r^2 + \omega^2 M^2 R_r R_t^{-1}} + R_r) + \omega^2 M^2} \tag{8}$$

When $R_r = Z_{eq}$, equation (5) can be simplified as:

$$\boldsymbol{U}_\mathbf{d} = M\boldsymbol{U}_{in} L_1^{-1} \tag{9}$$

Therefore, the WPT system under LCC/S compensation topology has constant voltage output characteristics. When the filter circuit is composed of capacitors as shown in figure 1, the relationship between output voltage and input voltage is $U_{dc} = \frac{\pi}{4} U_d$ (U_d is the amplitude in equation (9)). Then, the expression of output voltage U_{dc} can be deduced as:

$$U_{dc} = \pi M U_{in} \left(\sqrt{8} L_1 \right)^{-1} \tag{10}$$

where U_{in} is the effective value of high-frequency sine voltage in figure 1. Finally, the expression of the output voltage U_L to the load is obtained.

$$U_L = \pi \alpha M U_{in} \left(\sqrt{8}(1-\alpha) L_1 \right)^{-1} \tag{11}$$

From equation (11), we can get the expression of mutual inductance as:

$$M = \sqrt{8}(1-\alpha) L_1 U_L \left(\pi \alpha U_{in} \right)^{-1} \tag{12}$$

Therefore, real-time mutual inductance values can be obtained from the values of input voltage, real-time load voltage and duty cycle. From equation (11), the output power can also be expressed as:

IOP Conf. Series: Materials Science and Engineering **752** (2020) 012003 doi:10.1088/1757-899X/752/1/012003

$$P_{\mathrm{L}}=\frac{\pi^{2}\alpha^{2}M^{2}U_{\mathrm{in}}^{2}}{8(1-\alpha)^{2}L_{1}^{2}R_{\mathrm{L}}} \tag{13}$$

With the duty ratio α unchanged, the relation curve between output power and mutual inductance and load resistance is drawn by equation (13), as shown in figure 4. The output power decreases with the increase of load resistance and increases with the increase of mutual inductance. The output power fluctuates when the load equivalent impedance changes or the receiver coil is offset from the transmitter coil. For the battery load, the battery life will be reduced when the charging power fluctuates greatly.

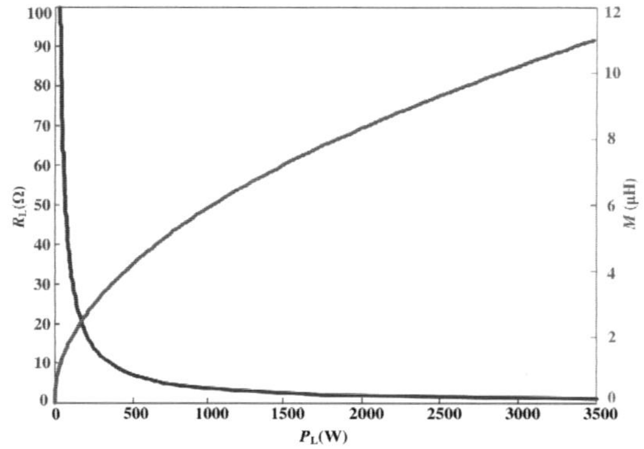

Figure 4. Curve of output power with mutual inductance and load resistance.

The relationship between duty ratio α and demand power P_{d} can be expressed as:

$$\alpha=\left(\pi MU_{\mathrm{in}}\left(\sqrt{8P_{\mathrm{d}}R_{\mathrm{L}}L_{1}^{2}}\right)^{-1}+1\right)^{-1} \tag{14}$$

The load equivalent impedance can be obtained from the load voltage $U_{\text{L-S}}$ and current $I_{\text{L-S}}$ obtained by real-time sampling. Set the overall efficiency of DC/DC circuit and rectifier circuit is η_{d}. Considering η_{d} and η, the equation (14) can be expressed as:

$$\alpha=\left(\pi MU_{\mathrm{in}}\left(\sqrt{8P_{\mathrm{d}}U_{\text{L-s}}L_{1}^{2}/(\eta_{\mathrm{d}}\eta I_{\text{L-s}})}\right)^{-1}+1\right)^{-1} \tag{15}$$

The output power of the system can be theoretically consistent with the load demand power by adjusting the value of α when the load resistance or mutual inductance changes, as shown in figure 5.

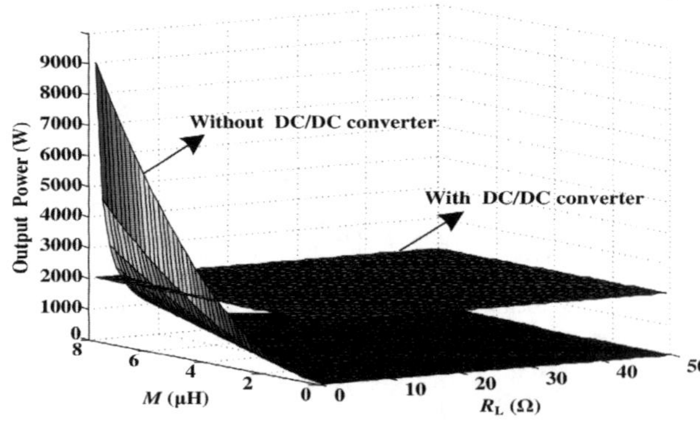

Figure 5. 3D diagram of output power with load resistance and mutual inductance.

3. Power stability control strategy

In this paper, the power stability control method only regulates in the secondary side and does not require real-time communication between the primary side and the secondary side. The goal of power stability can be achieved by detecting the load voltage and current and then controlling the DC/DC circuit. The system can output relatively stable power even if the coil position has a large deviation. So the power stability control method is very suitable for applications with a wide deviation range.

It is assumed that the power source voltage U_{in}, the internal resistance R_t of transmitting coil, the internal resistance R_r of receiving coil, the system working frequency f_0, the load demand power P_d, and the initial value of duty ratio α of DC/DC circuit are known. According to the load voltage $U_{L\text{-}S}$ and current $I_{L\text{-}S}$ detected in real time, the real-time mutual inductance value and the duty ratio of response load power demand can be obtained by equation (15). Therefore, this paper proposes a power stability control strategy by adjusting the duty ratio of DC/DC circuit based on load voltage and current monitoring. The control flow chart is shown in figure 6, and the specific process is as follows:

- Step 1. The value of U_{in}, L_1, R_t, R_r, η_d, f_o, P_d, initial duty ratio α(t) were determined, and stored to the controller.
- Step 2. Initialize the relevant constraints, including the minimum charge voltage U_{Lmin} and the maximum charging voltage U_{Lmax}, adjustment range of duty ratio ($\alpha_{min} \leq \alpha \leq \alpha_{max}$), and the minimum transmission efficiency η_{min}.
- Step 3. According to equations (7) and (12), the maximum and minimum values of mutual inductance M_{max} and M_{min} can be obtained. These are the constraint conditions whether the proposed control strategy can be implemented or not.
- Step 4. Real-time voltage value $U_{L\text{-}S}$ and current value $I_{L\text{-}S}$ of the load were collected, and real-time mutual inductance value was obtained according to equation (12).
- Step 5. Determine whether the mutual inductance value calculated by Step 4 satisfies the relation of $M_{min} \leq M \leq M_{max}$. If meet constraint conditions, according to the equation (6) to get the transmission efficiency of the η(t). If not, it indicates that stable power control cannot be achieved at this time, and the user is informed to turn off the system to stop charging.
- Step 6. According to the transmission efficiency η(t) from step 5, the adjustment value of duty cycle α(t+1) can be obtained from equation (15).
- Step 7. Set α(t)=α(t+1) according to the duty cycle adjustment value obtained by Step 6, and then output the value to switch S_t of DC/DC circuit. Meanwhile, go to Step 4 to conduct the control process at the next moment.

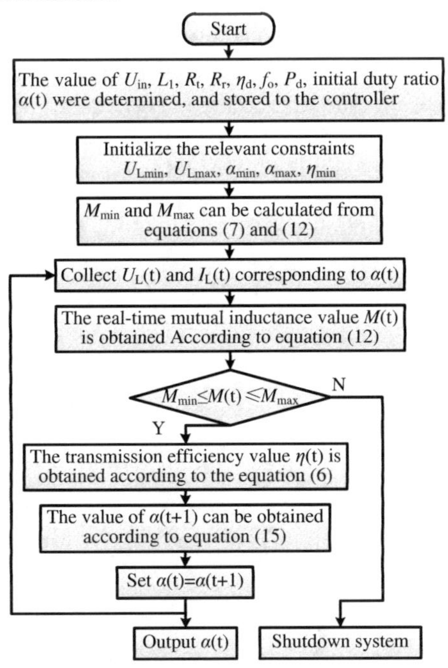

Figure 6. Flow chart of power stability control method.

4. Experimental and simulation analysis

In order to prove the feasibility of the proposed control strategy, this paper designs an experimental platform of WPT system based on the compensation topology of LCC/S and the secondary side contains DC/DC circuit according to the system architecture in figure 1, as shown in figure 7.

Set the input voltage U_{in}=200V, and the load demand power P_d is 1 kW. The input of the experimental platform is a DC voltage source, which is converted into a high-frequency square wave through the full-bridge inverter circuit. Since the input voltage $\boldsymbol{U}_{in}$ is sinusoidal wave, the voltage U_{DC} of the DC source should satisfy the equation of $U_{DC} = \pi U_{in}(2\sqrt{2})^{-1}$ if the square wave is equivalent to it, so the DC source voltage U_{DC} =222V. The full-bridge inverter circuit is composed of four SiC devices of IXFN70N120SK, and its working frequency is 85 kHz. The full-bridge rectifier circuit is composed of quick recovery diodes of DSEI2X101-12A and the Buck-Boost switch are IRFP460.

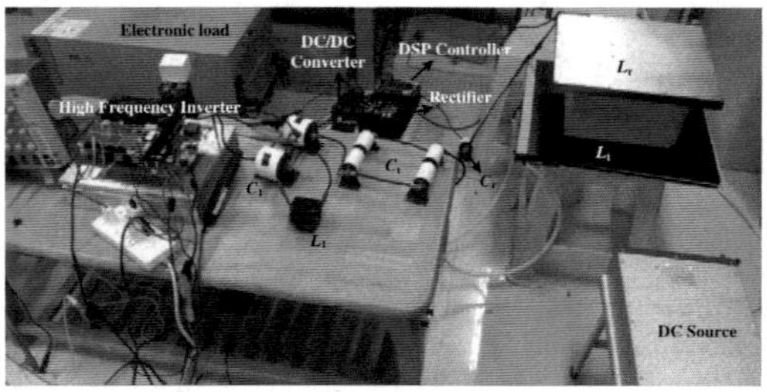

Figure 7. Experimental platform of WPT system.

The vertical distance between transmitter coil and receiver coil is 15 cm, and its dimensions are 46cm*46cm and 36cm*36cm respectively. The parameters of coils and the components in the LCC/S compensation topology are shown in table 1.

Table 1. Parameters of coils and components in the LCC/S compensation topology.

Parameter	L_1(μH)	C_1(nF)	L_t(μH)	R_t(Ω)	C_t(nF)	L_r(μH)	R_r(Ω)	C_r(nF)
Value	28	125.2	224.6	0.475	17.83	129.6	0.23	27.05

In order to prove that the proposed control strategy can effectively solve the problem of unstable power of the coil in the case of offset, it is necessary to test the changing trend of output power when the receiving coil and transmitting coil are offset longitudinally. When the load resistance is 10 Ω, set η_d = 0.97. Through the method of simulation and experiment, output power data with longitudinal offset from -16 cm to 16 cm were recorded, and the curve was drawn as shown in figure 8.

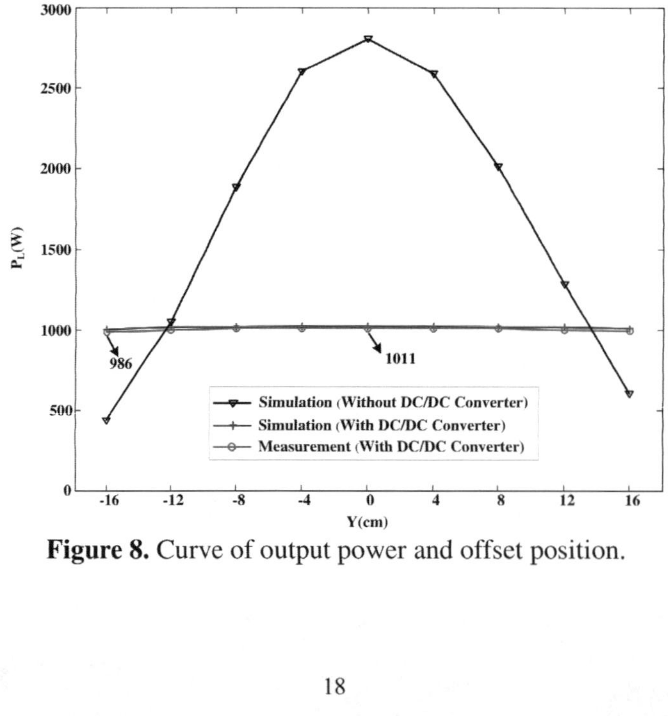

Figure 8. Curve of output power and offset position.

As can be seen from figure 8, the output power fluctuates greatly without DC/DC circuit when the longitudinal offset is from -16 cm to 16 cm. When the power stability control strategy proposed in this paper is adopted, the minimum value and maximum value of the output power are 986 W and 1011 W respectively. The fluctuation rate is about ±2%, which proves the effectiveness of the proposed control strategy.

The equivalent impedance will change during the charging process for battery loads. It is necessary to verify whether the load with different resistance values can guarantee the stability of output power under the control strategy proposed in this paper. Keep the receiving coil and transmitting coil without deviation, record the output power data corresponding simulation and experiment when the load resistance from 4 Ω to 20 Ω changes, drawn into a curve as shown in figure 9.

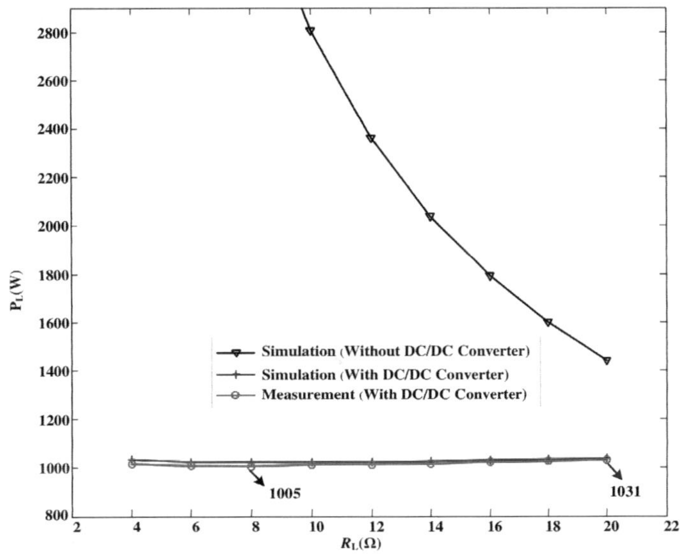

Figure 9. Curve of output power and load resistance.

The figure 9 shows that output power fluctuation is very big if there is no DC/DC circuit in the system when the load resistance changes, and the value of output power is very high when the load resistance value is less than 8 Ω, so the black curve in figure 9 shows only the load resistance is greater than 8 Ω corresponding output power values. The minimum value and maximum value of the output power are 1005 W and 1031 W respectively when the power stability control strategy proposed in this paper is adopted. The small variation range proves that the proposed control strategy has strong power stabilization effect for the load whose equivalent impedance will change.

5. Conclusions

This paper analyzes the WPT system with LCC/S compensation topology and the secondary side containing DC/DC circuit. The relationship between mutual inductance and load voltage and current, the relationship between mutual inductance and transmission efficiency, and the relationship between DC/DC circuit duty ratio and load voltage and current are deduced. The output power stability control strategy by detecting the load voltage and current and then controlling the DC/DC circuit is proposed. Finally, the WPT system with 1 kW output power is built through simulation and experiment, and the stability of output power after coils deviation and load resistance change under the control strategy proposed in this paper is analyzed. The simulation and experimental results verify the correctness and effectiveness of the theoretical analysis and the proposed control strategy.

Acknowledgments

This work was supported by National Key Research and Development funded Project (2018YFB0106300).

References

[1] Mou X, Groling O and Sun H 2017 Energy-efficient and adaptive design for wireless power transfer in electric vehicles *IEEE Transactions on Industrial Electronics* **64** 7250–7260.

[2] Liu H, Huang X, Tan L, Guo J and Wang W 2018 Dynamic wireless charging for inspection robots based on decentralized energy pickup structure *IEEE Transactions on Industrial Informatics* **14** 1786-1797.

[3] Basar M, Ahmad M, Cho J and Ibrahim F 2017 Stable and highefficiency wireless power transfer system for robotic capsule using a modified Helmholtz coil *IEEE Transactions on Industrial Electronics* **64** 1113-1122.

[4] Zhu Q, Wang L, Guo Y and Liao C 2016 Applying LCC compensation network to dynamic wireless EV charging system *IEEE Transactions on Industrial Electronics* **63** 6557-6567.

[5] Li S, Liu Z, Zhao H, Zhu L, Shuai C and Chen Z 2016 Wireless power transfer by electric field resonance and its application in dynamic charging *IEEE Transactions on Industrial Electronics* **63** 6602–6612.

[6] Choi S, Huh J, Lee W and Rim C 2014 Asymmetric coil sets for wireless stationary EV chargers with large lateral tolerance by dominant field analysis *IEEE Transactions on Power Electronics* **29** 6406-6420.

[7] Kan T, Lu F, Nguyen T, Mercier P and Mi C 2018 Integrated coil design for EV wireless charging systems using LCC compensation topology *IEEE Transactions on Power Electronics* **33** 9231-9241.

[8] Tan L, Zhang M and Wang S 2019 The Design and optimization of a wireless power transfer system allowing random access for multiple loads *Energies* **12**.

[9] Liu F, Yang Y, Jiang D, Ruan X and Chen X 2017 Modeling and optimization of magnetically coupled resonant wireless power transfer system with varying spatial scales *IEEE Transactions on Power Electronics* **32** 3240-3250.

[10] Wang A, Liu J and Wang X 2019 The load estimation and power tracking integrated control strategy for dual-sides controlled LCC compensated wireless charging system *IEEE Access* **7** 75749-75761.

[11] Tan L, Guo J and Huang X 2018 Coordinated source control for output power stabilization and efficiency optimization in WPT systems *IEEE Transactions on Power Electronics* **33** 3613-3621.

[12] Yao Y, Wang Y, Liu X, Lu K and Xu D 2018 Analysis and design of an S/SP compensated IPT system to minimize output voltage fluctuation versus coupling coefficient and load variation *IEEE Transactions on Vehicular Technology* **67** 9262-9272.

[13] Yang L, Li X, Liu S, Xu Z, Cai C and Guo P 2019 Analysis and design of three-coil structure WPT system with constant output current and voltage for battery charging applications *IEEE Access* **7** 87334-87344.

IOP Conf. Series: Materials Science and Engineering **752** (2020) 012004 doi:10.1088/1757-899X/752/1/012004

A Scheduling Model for Wind Power Consumption Considering Source-Charge Coordination of Combined Heat and Power System in Low-Carbon Environment

Yifan Lang[1], Dan Xu[2], Shuyan Liu[1] and Linxian Hu[1]

[1]School of Electrical Engineering and Automation, Harbin Institute of Technology, Harbin 150001, Heilongjiang Province, China
[2]China Electric Power Research Institute, Beijing 100192, China

E-mail: linxian_hu@163.com

Abstract. Increasing the penetration and consumption rate of new energy sources will help reduce or control carbon dioxide emissions and better promote green and low-carbon development. Based on the establishment of deep peak-load regulation model of pure condensing thermal power units, the model for the expansion of combined heat and power plant adjustment range, interruptible electro-thermal flexible load model and combined heat and power units carbon emission model, this paper introduces a quota-based carbon emission trading mechanism and constructs a source-load coordination scheduling model of combined heat and power system that considers the characteristics of heating network and aims at minimizing the total cost of units operation and carbon emission trading. The total coal consumption and wind curtailment absorption effect in three cases are compared by cases study. The outcomes display that the model in this paper can coordinate the scheduling resources on both sides of source and load, further diminish the total coal consumption and improve wind power consumption capacity based on improving the flexibility of source and load side, and achieve wind power consumption in low-carbon environment.

1. Introduction

In the face of global climate change, reducing carbon dioxide emissions has become a top priority for all countries in the world. Developing new energy sources and reducing the coal consumption of traditional thermal power are effective methods. However, wind curtailment problem becomes more and more prominent in the northern China. The main reasons for the serious problem in the northern region are the constraint of "Power Determined by Heat" of combined heat and power (CHP) units and peak-load regulation range limitation. For the reasons, Reference [1] proposes a deep peak-load regulation model for thermal power units. Reference [2] considers the heat storage feature of the heating network. In [3], the demand response is introduced to take part in active peak-load regulation based on adding heat storage devices on the source side.

The above references do not take into account the carbon emissions. With the implementation of the national energy-saving scheduling policy, low-carbon has become one of the core indexes concerned by enterprises. In many models involving carbon emissions trading, most of them are only concentrated in power system [4-5]. Although the models of the CHP system in low-carbon environment are constructed in [6], the effects of heating network characteristics are not taken into account, and ignore the influence of CHP units on the total carbon emissions.

In view of the above research, this paper establishes the deep peak-load regulation model of pure condensing thermal power units, the model for the expansion of CHP plant adjustment range,

Content from this work may be used under the terms of the Creative Commons Attribution 3.0 licence. Any further distribution of this work must maintain attribution to the author(s) and the title of the work, journal citation and DOI.
Published under licence by IOP Publishing Ltd

interruptible electro-thermal flexible load model, and introduces the carbon emission trading model of thermal power units and CHP units based on quota. Then the paper establishes the source-load coordination scheduling model of CHP system in consideration of the features of heating network with the purpose of minimizing the total cost of units operation and carbon emission trading. Finally, the simulation analyses of three cases reveal that the proposed model can further decrease the overall coal consumption and wind curtailment rate on the basis of improving the flexibility of on both sides of source and load, and improve the wind power absorption level of the CHP system.

2. Source-load model of CHP system

2.1. Deep peak-load regulation model

Deep peak-load regulation reconstruction of thermal power units can improve source side flexibility, and the process can be divided into three stages: basic peak-load regulation, non-oil steady state combustion deep peak-load regulation and oil-assisted combustion deep peak-load regulation [7].

At basic peak-load regulation stage, only the coal consumption of units should be considered.

$$C^{\mathrm{OP}} = p^{\mathrm{coal}}\left(a_0 + a_1 P + a_2 P^2\right) \tag{1}$$

Where: C^{OP} is the coal consumption cost (Yuan); p^{coal} is the coal price (Yuan/t); a_0, a_1 and a_2 are coefficients of thermal power unit consumption characteristic; P is generation power (MW).

With the increase of peak-load regulation depth, the rotor loss increases during the non-oil steady state combustion period and the service life decreases [8]. Moreover, oil is required to support burning during the oil-assisted combustion period. Therefore, the operation cost C^{CON} (Yuan) of deep peak-load regulation at different stages can be indicated by the piecewise function in [9].

For the sake of enhancing the passion for deep peak-load regulation, the units participating in the deep peak-load regulation should be given subsidy, which can be expressed by Equation (2). Moreover, the subsidy shall be shared by thermal power units that are not referred to deep peak-load regulation and wind turbines, and the sharing cost can be expressed as Equation (3).

$$C^{\mathrm{SUB}} = \begin{cases} 0, P \geq P^{\min} \\ p^{\mathrm{SUB}}\left(P^{\min} - P\right), P < P^{\min} \end{cases} \tag{2}$$

Where: C^{SUB} is the deep peak-load regulation subsidy (Yuan); p^{SUB} is the subsidy price of the unit power below the minimum technical output (Yuan/(MW·h)); $P^{\min}$ is the lower power limit of thermal power unit (MW).

$$C^{\mathrm{SHARE}} = q^{\mathrm{SHARE}} C^{\mathrm{SUB}} \tag{3}$$

Where: C^{SHARE} is the sharing cost (Yuan), and q^{SHARE} is the sharing ratio, which is determined by the ratio of rated capacity of the unit to total rated capacity of all the thermal power units that are not involved in deep peak-load regulation and wind power units.

2.2. Model for the expansion of CHP plant adjustment range

The coal consumption cost of a CHP unit can be calculated by Equation (4):

$$C^{\mathrm{CHP}} = p^{\mathrm{coal}}\left(e_0 + e_1 P + e_2 D + e_3 P^2 + e_4 PD + e_5 D^2\right) \tag{4}$$

Where: C^{CHP} is the coal consumption cost (Yuan); D is steam extraction rate for heating (t/h); $e_0 \sim e_5$ are coefficients of consumption characteristic;

The relationship between steam extraction rate and heat output Q (GJ/h) of CHP unit is:

$$Q = D\Delta H / 1000 \tag{5}$$

Where: ΔH is vapour enthalpy drop (kJ/kg).

The most economical scheme for expanding the regulation range of CHP unit is to configure thermal storage tanks in CHP plants. Heat storage tanks store heat during non-wind curtailment period while release heat in wind curtailment period. As shown in Figure 1, the interval ABCD is the original electro-thermal output regulation range of the CHP unit. After the thermal storage tank is installed, the range extended to the interval AA'B'C'C"D'. In addition, P^{M} and P^{m} are the upper and lower power limit of CHP unit (MW), respectively; D^{M} and D^{m} are the upper and lower steam extraction rate limit of CHP unit (t/h), respectively.

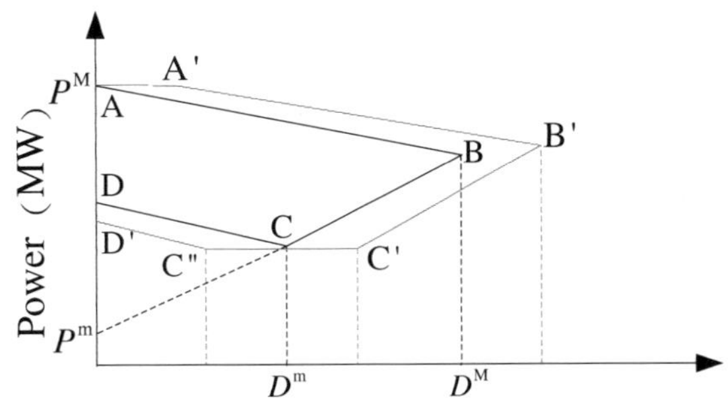

Figure 1. Schematic diagram of expanding regulation range with thermal storage tanks.

Set the heat storage at the initial time is E_0 (MW·h), and the relationship between heat storage and heat storage or release power can be indicated by Equation (6):

$$E_t = E_{t-1} + h_t^{\mathrm{v}} \tag{6}$$

Where: E_t is the heat storage at time t (MW·h); h_t^{v} is the heat storage or release power at time t (MW). This paper stipulates that the heat storage is positive and the heat release is negative.

The balance of heat storage and release should be satisfied in one cycle.

$$\sum_{t\in T} h_t^{\mathrm{v}} = 0 \tag{7}$$

Where: T is the total number of scheduling cycles (h).

2.3. Electro-thermal flexible load

Installation of electric boilers is one of the effective measures to consume wind curtailment. This paper assumes that the electric boilers are installed on the load side, and uses wind curtailment start-up and shut-down control [10]. The electric boilers work during the wind curtailment period, consuming power while heating, decreasing heat power of CHP units, and they are stop working during the non-wind curtailment period. Therefore, the electric boilers can be regarded as an electro-thermal flexible load that is interruptible and adjustable in size. The relationship between power P^e (MW) consumed by electric boiler and heat output Q^e (GJ/h) of electric boiler can be referred to [11].

Taking the power generation of the pure condensing units, the electro-thermal power of CHP units, and the power consumed by electric boilers as the optimization variables, which can make full use of the scheduling resources on both sides of the source and load, and realize the coordination of the source and load of the CHP system.

3. Heating network model

Heating network is partitioned into water supply network and return water network. The topological structure of thermal system is shown in Figure 2. T^s, T^o and T^r are the supply, return and return mixing temperature of node, respectively. m is the pipeline flow. m^{q} is the node injection water flow.

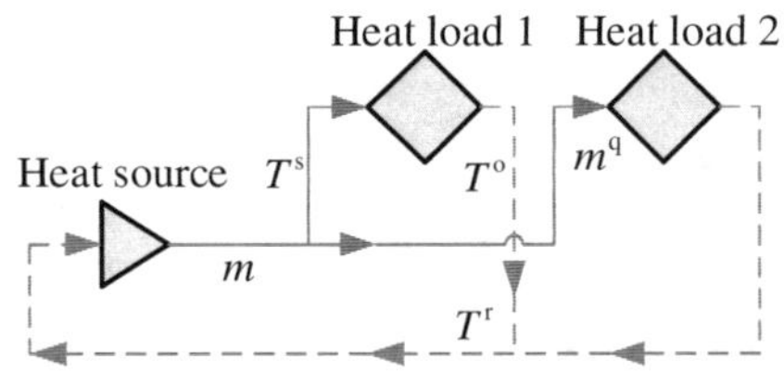

Figure 2. Topological structure diagram of thermal system.

Since the electric boilers are installed on the load side, the flow at the heat load node can be calculated according to Equation (8). And the temperature relationship between the water supply network node and the return water network node can be referred to [12].

$$m_j^q = (Q_j^{load} - Q_j^e) / C^P (T_j^s - T_j^o) \tag{8}$$

Where: Q_j^{load} is the load at heat load node j (GJ/h) and C^P is specific heat capacity (kJ·kg^{-1}·K^{-1}).

The heat loss in the process of hot water transportation is [12]:

$$\Delta Q_{ij} = 2\pi\lambda^d (T_{ij} - T^0) l_{ij} / \ln(D_{ij} / d_{ij}) \tag{9}$$

Where: T_{ij} and ΔQ_{ij} are the temperature of medium (K) and heat loss (GJ/(m·h)) of the pipeline ij, respectively. λ^d is the thermal conductivity of insulation layer of pipeline (GJ·m^{-1}·K^{-1}); T^0 is the surface temperature of insulation layer (K); l_{ij}, D_{ij} and d_{ij} are the length, external diameter and internal diameter of pipeline ij (m), respectively.

The time delay characteristic equation of the heating network is:

$$\tau_{ij} = \pi \rho l_{ij} \mathrm{d}_{ij}^2 / 4m_{ij} \tag{10}$$

Where: τ_{ij} is heat transmission time (h), and ρ is water density (kg/m^3).

4. Carbon emission trading model

A market model of global carbon emissions trading is a market based on quota, which determines transactions by the difference in size between actual emissions and quota-based emissions [13].

The cost of carbon emissions trading can be expressed as:

$$C^{carbon} = \sum_{t=1}^{T} p^{carbon} \left(E_t^c - E_t^q \right) \tag{11}$$

Where: C^{carbon} is the total cost of carbon emissions trading (Yuan); p^{carbon} is the carbon transaction price (Yuan/t); E_t^c is the actual carbon emission of units at time t (t); E_t^q is the quota-based carbon emission of units at time t (t).

Because the output of CHP unit includes electric energy and heat energy, it is necessary to convert heat into equivalent electric quantity when calculating the carbon emissions of CHP units.

Among them, the actual carbon emissions of the units are:

$$E_t^c = \sum_{j=1}^{S} \sigma_j p_{t,j}^{CON} + \sum_{i=1}^{R} \sum_{n=1}^{N_i} \sigma_n (P_{t,i,n}^{CHP} + Q_{t,i,n}^{CHP} / 3.6\beta) \tag{12}$$

Where: σ_j and σ_n are the carbon emission intensity of thermal power unit and CHP unit, respectively (t/MW·h); $p_{t,j}^{CON}$ is the power of thermal power unit j at time t (MW); S is the total number of thermal power units; N_i is the total number of CHP units in CHP plant i; $P_{t,i,n}^{CHP}$ and $Q_{t,i,n}^{CHP}$ are the power (MW) and heat power (GJ/h) of CHP unit n in CHP plant i at time t, respectively; β is thermo-electric ratio; R is the number of CHP plants.

The carbon emission quota amount of units can be expressed as:

$$E_t^{\mathrm{q}} = q\left[\sum_{j=1}^{S} P_{t,j}^{\mathrm{CON}} + \sum_{i=1}^{R}\sum_{n=1}^{N_i}(P_{t,i,n}^{\mathrm{CHP}} + Q_{t,i,n}^{\mathrm{CHP}} / 3.6\beta)\right] \tag{13}$$

Where: q is the power carbon emission quota (t/MW·h), which is determined by the "regional grid baseline emission factor".

5. Scheduling model of CHP system

5.1. Objective function

The goal is to minimize the overall operating cost and carbon emission transaction cost of all units. Furthermore, the operating cost of pure condensing thermal power units includes not only the basic coal consumption cost, but also the subsidy and the allocation cost of units.

$$\min F = \sum_{t=1}^{T}\left[\sum_{i=1}^{R}\sum_{n=1}^{N_i} C_{t,i,n}^{\mathrm{CHP}} + \sum_{j=1}^{S}\left(C_{t,j}^{\mathrm{CON}} - C_{t,j}^{\mathrm{SUB}} + C_{t,j}^{\mathrm{SHARE}}\right)\right] + C^{\mathrm{carbon}} \tag{14}$$

Where: F is the total cost (Yuan).

5.2. Constraints

Power balance constraints, district heat balance constraints, unit ramping constrains, upper and lower limits of unit output constraints, electric boiler capacity constrains, thermal storage tanks capacity constraints and heat storage or release power constraints are all calculated according to [14].

Heat output of heat source nodes constraints can be expressed as:

$$Q_{t,j}^{\mathrm{CHP}} - \sum_{m=1}^{M} 3.6 h_{t,i,m}^{\mathrm{v}} = C^{\mathrm{P}} m_j^{\mathrm{CHP,q}} (T_{t,j}^{\mathrm{CHP,s}} - T_{t,j}^{\mathrm{CHP,o}}) \tag{15}$$

Where: $Q_{t,j}^{\mathrm{CHP}}$, $T_{t,j}^{\mathrm{CHP,s}}$ and $T_{t,j}^{\mathrm{CHP,o}}$ are the heat output (GJ/h), the supply temperature (K) and the return temperature (K) of the heat source node j at period t, respectively; $m_j^{\mathrm{CHP,q}}$ is the flow of the heat source node j (10^6 kg/h); M is the total number of thermal storage tanks.

6. Case study

6.1. Case system

The case system composes of two thermal power plants, two CHP plants and one wind farm. Each thermal power plant contains three pure condensing thermal power units. The CHP plant 1 and 2 are equipped with three and four CHP units respectively. The installed capacity of the wind farm is 400 MW. Each CHP plant has three thermal storage tanks and three heat-exchanger stations. Assuming that only one thermal power unit in each thermal power plant takes part in the deep peak-load regulation transformation. The carbon emission intensity of each unit is shown in Table A1. Other parameters involved in the model are shown in Table A2. Take one day as a scheduling cycle and one hour as a scheduling period. Heating network structure and the parameters of pipeline are shown in [14]. The consumption characteristics coefficients, operation parameters of each unit, electric boiler capacity, power load data, heat load data and wind power prediction data are all mentioned in [15].

6.2. Study of wind power absorption effect

In this paper, the following three cases are simulated and researched.

Case 1: Consider only source-load coordination

Case 2: Source-load coordination considering carbon emissions

Case 3: Source-load coordination considering carbon emissions and heating network characteristics

Total coal consumption and wind curtailment rate in three cases are shown in Table 1. Figure 3 is the curve of power of thermal power plant, power consumption of electric boiler and capacity of thermal storage tanks in case 1. The heat output curves of three cases are illustrated in Figure 4. Figure

5 and Figure 6 are the power of thermal power units 3-1, 4-3 and CHP units 1-1, 2-3 in case 1 and case 2, respectively.

Table 1. Total coal consumption and wind curtailment rate in each case.

case	Total coal consumption (t)	wind curtailment rate
case 1	14938.12915	4.620%
case 2	14912.27582	4.619%
case 3	14392.04154	2.306%

Comparing the data of case 1 and case 2 in Table 1, we can see that the total coal consumption decreases and wind curtailment rate decreases slightly after the introduction of carbon emission transaction cost into the objective function, while wind curtailment rate of case 3 decreases greatly when the characteristics of the heating network are considered in the model. Combining with Figure 4, it can be seen that the heat output of the CHP unit decreases greatly in case 3, which further reduces the total coal consumption.

From Figure 3 and Figure 4, it can be pointed out that the deep peak-load regulation in case 1 makes the power of thermal power plant relatively low. Moreover, during the wind curtailment period, the thermal storage tank releases heat and the electric boiler works; during the non-wind curtailment period, the thermal storage tank stores heat and the electric boiler shuts down, which reduces the heat output of CHP plants. Therefore, source-load coordination can improve the flexibility of the system, thereby enhancing the wind power absorption ability.

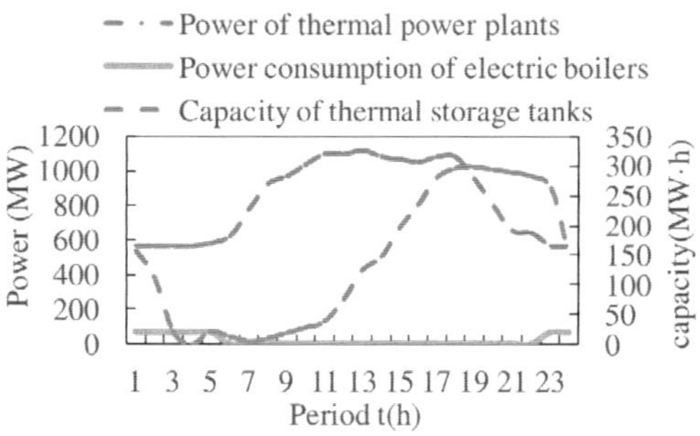

Figure 3. Power of thermal power plants, power consumption of electric boilers and thermal storage tanks capacity in case 1.

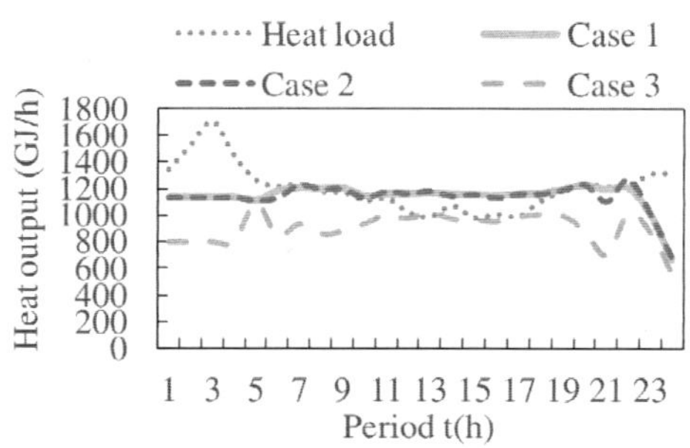

Figure 4. Heat output of CHP plants in 3 cases.

Furthermore, comparing the heat output curves of the CHP plants in case 2 and case 3 in Figure 4, it can be seen that under the influence of the time delay characteristics, the curve of case 3 is shifted to the left, and when the model takes into account the characteristics of heating network, the heat output of the CHP plant decreases dramatically, which further enlarges the space for wind power into grid. Therefore, the model in this paper can further improve the wind power absorption rate and alleviate the peak-load regulation pressure of grid more effectively.

Since the carbon emission trading mechanism of thermal power units and CHP units is considered in case 2, the system will give priority to reducing the output of units with higher carbon emission intensity and increase the output of units with lower carbon emission intensity, in order to attain the goal of low-carbon. Therefore, in Figure 5 and Figure 6, it is clear that the power of thermal power unit 3-1 (σ_{3-1}=1.20 t/MW·h) and CHP unit 1-1 (σ_{1-1}=1.20 t/MW·h) decrease significantly, while the power of thermal power unit 4-3 (σ_{4-3}=0.42 t/MW·h) and CHP unit 2-3 (σ_{2-3}=0.42 t/MW·h) increase slightly.

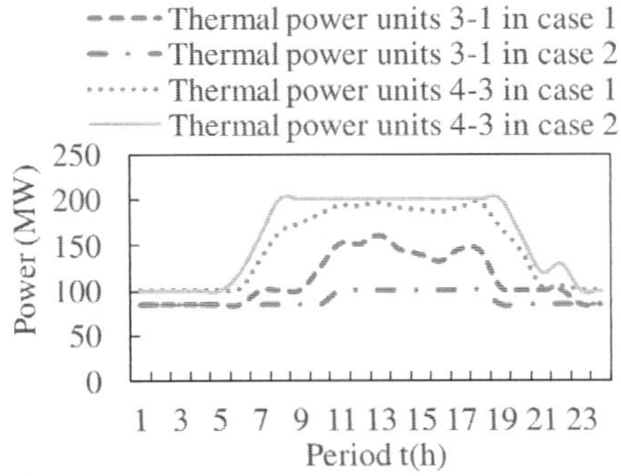

Figure 5. Power of thermal power unit 3-1 and 4-3 in case 1 and case 2.

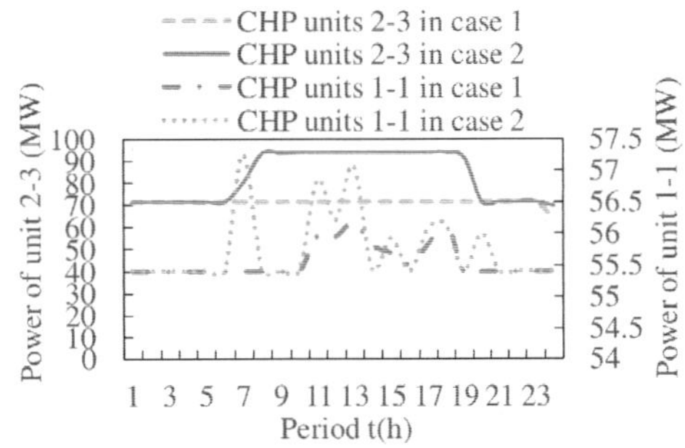

Figure 6. Power of CHP unit 1-1, 2-3 in case 1 and case 2.

7. Conclusions

A dispatch model for wind power consumption considering source-charge coordination of CHP system in low-carbon environment is proposed in this paper. With the introduction of carbon emission transaction cost into the objective function, the output of units with higher carbon emission intensity decrease, and the total coal consumption of the system decreases accordingly. Moreover, the characteristics of heating network make the heat output curve of CHP units shift, and the peak value of the heat output decreases. In addition, the simulation analyses of various cases shows that the model of this paper can further reduce the total coal consumption, improve the wind power consumption level based on improving the flexibility of the source-load side, and achieve wind power absorption in low-carbon environment.

8. Appendices

Table A1. Carbon emission intensity of each unit.

Unit number	1-1	1-2	1-3	2-1	2-2	2-3	2-4	3-1	3-2	3-3	4-1	4-2	4-3
σ_n / σ_j (t/MW·h)	1.20	0.96	1.12	1.20	0.98	0.42	0.50	1.20	0.96	1.12	0.40	0.98	0.42

Table A2. Other parameters.

Parameter	ΔH	p^{coal}	$p_{\mathrm{B}}^{\mathrm{SU}}$	C^{P}	ρ	p^{carbon}	β	q	E_0
Value	2327.53	680	72	4.182	1000	74.2	0.8	0.789	25

Acknowledgments

This work is supported by the National Natural Science Foundation of China (51877049).

References

[1] Lin L *et al* 2017 Multi-angle economic analysis of deep peak-load regulation of thermal power units under large-scale wind power grid connection *Autom. Elec. Power Syst.* **41** 21–7
[2] Bi Q S *et al* 2014 Research on deep peak-load regulation of large heating unit based on heat storage characteristics of heating network and building *Turbine tech.* **56** 141–4
[3] Cui X *et al* 2018 An optimal dispatch with source-load coordination of CHP system considering wind power consumption *Elec. Power Autom. Equip.***38** 74–81
[4] Wang S Y *et al* 2019 An optimal study on low-carbon power generation of power system considering deep peak-load regulation of thermal power *Global Energy Internet* **2** 226–31
[5] Chen L *et al* 2016 An optimized dispatch model of wind power integration considering carbon emissions trading and demand response *Journal of NCEPU (Natural Sci. Edn.)* **43** 104–10

[6] Zou C L *et al* 2019 Optimal scheduling of CHP system considering flexible load and electric boiler in low-carbon environment *Elec. Meas. and Inst.* 1–8

[7] Lin L *et al* 2017 Economic dispatch and benefit analysis of power system based on d deep peak-load regulation of thermal power classification *Power Syst. Tech.* **41** 2255–63

[8] Wu R K *et al* 2018 Effect of deep peak-load regulation on steam turbine equipment of coal-fired units *Thermal Power Gen.* **47** 89–94

[9] Deng T T *et al* 2019 Optimal dispatch of wind power system including demand response and deep peak-load regulation of thermal power *Autom. Elec. Power Syst.* **43** 1–8

[10] Deng J L *et al* 2016 Research on multi-time scale rolling scheduling strategy for CHP system *Power grid tech.* **40** 3796–802

[11] Li J J *et al* 2015 Study on wind curtailment consumption scheme based on electric boilers for peak-load regulation at secondary heating network *Power Syst. Tech.* **39** 3286–91

[12] Shao S Q *et al* 2018 A scheduling method of CHP system considering heating network characteristics *Power Syst. Prot. Control* **46** 24–30

[13] Lou S H *et al* 2014 Optimal dispatch of power system with large-scale photovoltaic power in carbon trading environment *Elec. Power Syst.* **38** 91–7

[14] Guo F H *et al* 2018 A scheduling model for wind curtailment consumption based on storage electric boilers for peak-load regulation at secondary heating network *Autom. Elec. Power Syst.* **42** 50-6

[15] Deng J L *et al* 2016 Mechanism and economic analysis of wind curtailment consumption with electric boilers for peak-load regulation at secondary heating network *Autom. Elec. Power Syst.* **40** 41–7

IOP Conf. Series: Materials Science and Engineering **752** (2020) 012005 doi:10.1088/1757-899X/752/1/012005

A Numerical Relay Implementation for Overcurrent Protection Based on ARM Cortex – M4 Microprocessor

Anh V Nguyen[1,a], Lien B Nguyen[1,b] and Anh Hoang[1,c]

[1]School of Electrical Engineering, Hanoi University of Science and Technology, Hanoi, Vietnam

[a]anh.nguyenvan1@hust.edu.vn; [b]Lien.nguyenbich@hust.edu.vn; [c]Anh.hoang@hust.edu.vn

Abstract. Numerical overcurrent protection relays are the most extensively used devices to safeguard the power system from detrimental faults which might lead to a catastrophe. This paper presents a practical implementation of a numerical overcurrent protection relay based on ARM Cortex – M4 microcontroller embedded with a floating-point unit which has ability to perform complex calculation operations with a great accuracy for real-time protection algorithms. In order to implement, the system is designed to capture analog values of input currents by using Hall current sensors, analog conditioning circuits and 12 – bit A/D converters. Based on the measured input current values, a real-time signal processing algorithm for overcurrent protection is applied to produce an inverse - time characteristic according to the IEEE Standard C37.112-1996. The experimental results demonstrate that our numerical relay produces great performances in the real-time and has ability to protect electrical components from exceeded current faults satisfying the IEEE standard for overcurrent relays.

1. Introduction

The most crucial purpose of a power system is to generate and distribute electric energy to consumers without interruptions. However, the power system is always vulnerable to faults due to either natural disasters or by misoperation of the system [1]. This can result in permanent damage to costly power system elements and disruptions of power supply to customers. In order to assist the power system to sustain faults as well as minimize the damage in critical components, overcurrent protection relays have been considerably utilized to protect the system from sudden rise in current.

An overcurrent relay uses inverse-time characteristics to determine tripping time to protect electrical elements from the power abnormal operating conditions. With this operating characteristic, the relay gives a shorter tripping time for high current magnitudes than that of low current magnitudes. There are two different types of relays: electromechanical relays and microprocessor-based or numerical relays. Compared to electromechanical relays, numerical protection ones are advantageous on the account that they provide better protection, reliability, troubleshooting, and fault information. This can explain a reason why electromechanical relays are gradually replaced by numerical ones in the power system [2, 3].

There are a number of reseachers who are interested in investing advanced inverse-time characteristics protection algorithms for numerical relays. Basically, there are two main methods: Direct interpolation from stored data [4, 5] and use of data-fitting formulas [6]. Direct interpolation

Content from this work may be used under the terms of the Creative Commons Attribution 3.0 licence. Any further distribution of this work must maintain attribution to the author(s) and the title of the work, journal citation and DOI.
Published under licence by IOP Publishing Ltd

methods rely on storing tripping time data in the memory, and then selecting an appropriate tripping data point based on the ratio between an input current and a pickup current. This method is not accurate because most of the time, the actual tripping time is not match with the one stored in the memory, then an interpolation is neccessary to determine the approximated time. On the other hand, data fitting fomulas directly use the input current to determine a tripping time based on the inverse – time fomulus, thus, this method provides more accurate tripping time results for the protection relays.

This paper reports an implementation of a numerical relay for overcurrent protection based on ARM cortex – M4 microcontroller based on the data fitting formulas method. Embedded many advanced signal processing features, this 32-bit microcontroller is able to perform a complex protection algorithm efficiently as well as communicate with other devices easily. Besides, this paper also focuses on an overcurrent protection algorithm in order to produce an operating characteristic required by the IEEE Standard [7].The rest of the paper will be organized as follows. The section 2 presents a theory of overcurrent protection based on the IEEE Standard. Next, the section III provides an overview of the system including both hardware design and software design. Based on the section 2 and section 3, the section 4 details an overcurrent protection algorithm in order to achieve a desired inverse – time characteristic. Finally, experimental results and conclusions will be made in the section 5 and section 6.

2. Overcurrent Protection

Basically, an inverse – time current characteristic is defined by two regions: an operation (trip) region and a rest region. The operation region is applied for the condition where the input current (I) is greater than the pickup value (I_p). A relationship between tripping time (t (I)) and input current is determined by

$$\begin{cases} t(I) = \dfrac{A}{M^p - 1} + B \\ M = \dfrac{I}{I_p} > 1 \end{cases} \tag{1}$$

Where A, B and p are constants and selected for a specific inverse – time characteristic. The IEEE standard [7] specifies three types of characteristics: moderately inverse, very inverse, or extremely inverse corresponding to their constants shown in the table 1. The rest region, on the other hand, is defined for the condition where the input current is smaller than the pickup current. The tripping time in this region is calculated by

Table 1. Standard constants regarding to the types of the inverse time-current characteristic.

Characteristic	A	B	p	T_r
Moderately inverse	0.052	0.114	0.02	4.85
Very inverse	19.61	0.491	2	21.6
Extremely inverse	28.2	0.122	2	29.1

$$\begin{cases} t(I) = \dfrac{T_r}{M^2 - 1} \\ M = \dfrac{I}{I_p}\ ; 0 < M < 1 \end{cases} \tag{2}$$

Where T_r is a constant depending on the type of the characteristic as shown in the table 1.

From the equation (1) and (2), it can be realize that the tripping time has positive values in the operating region while having negative values in the rest region. To combine both regions, the following dynamic equation must be satisfied

$$\int_0^{T_0} \frac{1}{t(I)} dt = 1 \tag{3}$$

Where T_0 is total tripping time when the current changes during the fault. The equation (3) implies that the tripping time for a time-varying input is equal to the harmonic mean of the delays relative to the current segments considered as constant over a very short time interval. In a discrete form, the above integral equation can take a form of a sum with integral step Δt as

$$\begin{cases} \sum_{i=0}^{N} \frac{\Delta t}{t_i(I)} = 1 \\ T_0 = N.\Delta t \end{cases} \tag{4}$$

3. System Overview

A completed hardware description of our developed numerical overcurrent relay is presented in the figure 1.

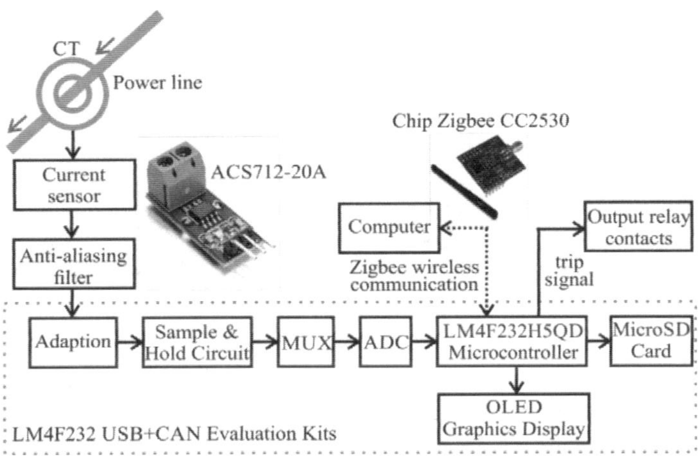

Figure 1. The numerical overcurrent relay hardware overview

The Allegro ACS 712 – 20A, based on the Hall Effect to sense either AC or DC current, are used to measure the input current. This sensor allows a linear conversion of the current margin [-20A, 20A] to the voltages [0.5V, 4.5V]. The voltage signal then is passed through an anti – aliasing second order Butterworth filter with a cutoff frequency of 500 Hz in order to attenuate electrical noises which have frequencies with 500 Hz or higher. It is necessary to avoid aliasing while sampling the input signals.

The center of this numerical relay is based on Stellaris LM4F232 evaluation board from Texas Instrument [8]. The figure 1 shows some parts which are already built in the evaluation board and utilized for this project. A purpose of the adaptation is to scale down the filtered voltage range into the 0 – 3V range of the 12 – bit ADC of the microcontroller. After receiving and processing signals from the ADCs, the microcontroller displays the results onto the 96 x 64 color OLED screen using SSP2 peripheral. In addition, it is also communicated with a host computer via either USB cable or Zigbee wireless communication using UART protocol. The relay obtains setting parameters from the computer for its overcurrent protected algorithm and send processed data to the computer for monitoring and recording. In a case of tripping, an output pin of the LM4F232H5QD chip emits a control signal to an electromechanical relay with optocoupler isolation in order to change the status of contacts which is used for a circuit breaker to disrupt the power supply circuit.

4. Algorithm

As discussed in the section 2, the inverse – time current characteristic of a relay is determined by the tripping time using the equation (1) and (2). However, before applying these equations, a RMS value of the input current must be found first. Measuring a RMS value of a current can be done by either time domain algorithms or frequency domain algorithms. In this project, the method of average values [4] is applied in order to estimate the RMS value of the input current because this method consume less RAM memory and its convergence is much faster than others. When applied this method, the basic assumption is that the signals are sinusoidal. An average value of the current is calculated by

$$\begin{aligned} I_{avg} &= \frac{1}{T}\int_{t}^{t+T} |i(\mathrm{t})|\,dt = \frac{1}{T}\int_{t}^{t+T} |I_m \sin(\omega \mathrm{t}+\varphi)|\,dt \\ &= \frac{2}{T}\int_{t}^{t+T/2} |I_m \sin(\omega \mathrm{t}+\varphi)|\,dt = \frac{2I_m}{\pi} \end{aligned} \tag{5}$$

Where Im and T is a peak value and a period of the current.

Based on the equation (5) and the definition of the RMS value of the current, the relationship between the average value and RMS value of the current can be found as following

$$\begin{aligned} I_{RMS} &= \sqrt{\frac{1}{T}\int_{t}^{t+T} i^2(\mathrm{t})dt} = \sqrt{\frac{1}{T}\int_{t}^{t+T} I_m^2 \sin^2(\omega \mathrm{t}+\varphi)t} \\ &= \frac{I_m}{\sqrt{2}} = \frac{\pi I_{avg}}{2\sqrt{2}} \end{aligned} \tag{6}$$

From the equation (6), RMS value of the current can be easily determined by estimating its average value. Let's assume that there are m samples of the current i(k) that are captured by the A/D converter within a period T. Then, the equation (6) is represented in a discrete form as below

$$\begin{aligned} I_{RMS} &= \frac{\pi I_{avg}}{2\sqrt{2}} = \frac{\pi}{2\sqrt{2}}\frac{1}{m}\sum_{k=1}^{m}|i(k)| \\ &= \frac{\pi}{m\sqrt{2}}\sum_{k=1}^{m/2}|i(k)| \end{aligned} \tag{7}$$

The equation (7) implies that the average method allows us to find out the RMS value of the current during a time of half period with simple operations including only addition and multiplication.

After IRMS found, the tripping time can be easily obtained. However, to deal with the dynamics of the current change, the tripping condition described in the equation (4) must be met. In digital implementation, this can be done by using an accumulation function C(I)

$$C_k(I) = \frac{T_c}{t_k(I)} + C_{k-1}(\mathrm{I}) \tag{8}$$

Where subscript k indicates a loop kth of the program, and Tc is a time constant representing for a time interval between two consecutive loops of the program. A value of the constant Tc is selected by considering the sampling rate of A/D converter as well as computing speed of operators.

Based on the value of tripping condition Ck(I), the microcontroller determines whether tripping signals are sent to change the contact status or not. The value of $C_k(I)$ tends to increase when overcurrent is present, and decrease as the current is below the pickup value I_p.

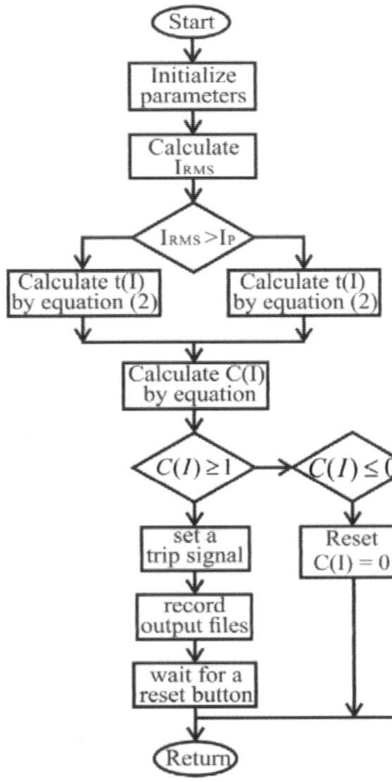

Figure 2. An algorithm for overcurrent protection

To summarize all above points, the figure 2 shows an overcurrent protection algorithm for our developed relay.

5. Experimental results

5.1 Setup

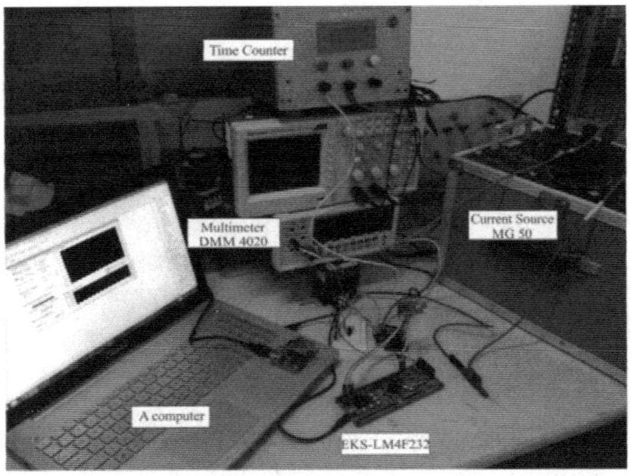

Figure 3. Experimental setup with unpacked parts of our numerical overcurrent relay

In order to investigate the performance of the developed system, several tests were conducted with our numerical overcurrent relay. The figure 3 presents an experimental setup for the evaluation tests. To generate test currents, we use a voltage – current set for protecting relay testing Cotel M50 which acts as a current source ranging up to 50A. A digital multimeter DMM 4020 of Tektronix is also recruited to measure the test current values during experiments. A computer is connected to a RF Zigbee DRF1605H module to wirelessly communicate with the relay. The user interface in the computer enable a user to:

1) Control data acquisition speed via the timer setup.

2) View the measured current and tripping condition sent from the relay and record them into files as desired.

3)Change defined protection modes (moderate, very and extreme as described by the IEEE standard) or manually change parameters of the inverse – time characteristics.

We also tested the communication between the relay and the host computer via Zigbee wireless. We found that they were still able to communicate with each other very well in few hundred meters away. Finally, a time counter connected with output contacts of the relay is used to measure the tripping time.

5.2. Tripping time

Table 2. An accuracy of moderately inverse – time characteristic obtained by our developed relay

Test current (A)	Calculated trip time (s)	Relay average trip time (s)	Error (%)
0.85	-	-	-
1.05	52.87	52.59	0.53
1.5	6.44	6.34	1.55
2	3.80	3.77	0.79
2.5	2.90	2.85	1.68
3	2.43	2.40	1.32
5	1.69	1.71	1.28
7	1.41	1.42	0.59
9	1.26	1.28	1.56
10	1.21	1.19	1.39

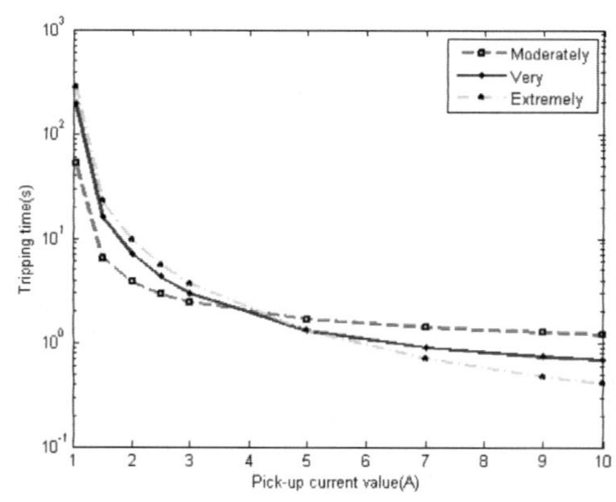

Figure 4. Inverse – time characteristics of our developed numerical overcurrent relay

We would like to measure the accuracy of the inverse – time characteristic and the difference between the protective characteristic types produced by our proposed relay. There are two experiments conducting for this purpose. The first experiment was carried out to measure an accuracy of moderately inverse - time characteristic of the developed system. In this experiment, the input current was varied to some preset values, and the tripping time of our numerical relay corresponding to a particular current value was recorded by the time counter. For the consistency, for each test current value, there are three repeated trials and the tripping time is averaged. Tripping time calculated by the formulas of the IEEE Standard C37.112-1996 was also provided for a comparison purpose as tabulated in the table 2. From the table 2, it can be easily seen that the moderately inverse – time characteristic obtained from our system was highly accurate. When comparing with values determined from the IEEE standard, a percentage error is less than 2% which is highly acceptable.

The second experiment was to measure three types (moderately, very and extremely) of the inverse –time characteristic produced by our numerical overcurrent relay. The figure 4 shows these characteristics on semi-logarithm graph. From this plot, we can see that there were obvious differences between protection modes of our relay and all these characteristics were highly consistent to those of the IEEE protection standard.

6. Conclusion

This paper presents an implementation of a numerical overcurrent relay based on ARM Cortex M4 microcontroller which has advanced features such as Super Harvard Architecture and Floating Point Processing Unit. The system uses a Hall current sensor ACS 712 – 20A to convert the input current into a voltage signal which is filtered, conditioned, and then sampled by 12-bit high speed A/D converter. ARM Cortex M4 combined with powerful signal processing functionality effectively performs the average method to estimate RMS value of current and the overcurrent protection algorithm. The experiment results prove that the protection characteristics produced by our proposed system are highly accurate and very consistent with those of the IEEE protection standard. In the future, we would like to apply some advanced protection techniques such as artificial neural network in order to improve the performance of this system.

Acknowledgements
This research is funded by Hanoi University of Science and Technology (HUST) under grand number is T2018 – PC - 051

References

[1] L. Wenchen, S. Lu, Z. Weimei, and L. Chunqing, "Power system reliability analysis system based on PSASP and fault enumeration method and applications," in Electricity Distribution (CICED), 2014 China International Conference on, 2014, pp. 1323-1326.
[2] W. Li, L. Sun, W. Zou, and C. Luo, "Power system reliability analysis system based on PSASP and fault enumeration method and applications," in 2014 China International Conference on Electricity Distribution (CICED), 2014, pp. 1323-1326.
[3] S. Richards, D. Chatrefou, and A. Procopiou, "Movement to the full digital substation for primary distribution," in 22nd International Conference and Exhibition on Electricity Distribution (CIRED 2013), 2013, pp. 1-4.
[4] Z. N. Stojanovic and M. B. Djuric, "Table Based Algorithm for Inverse-Time Overcurrent Relay," Journal of Electrical Engineering, vol. 65, 01/21 2014.
[5] G. V. Kshirsagar, G. Mulay, and S. Yeolekar, TMS320F28335 based single phase overcurrent protection Implementation using Numerical Relay, 2014.
[6] L. S. A. Armando Guzmán, and C. Labuschagne, "Adaptive Inverse-Time Elements Take Microprocessor-Based Technology Beyond Emulating Electromechanical Relays," 68th Annual Georgia Tech Protective Relaying Conference, Georgia, April 2014.
[7] G. Benmouyal, M. Meisinger, J. Burnworth, W. A. Elmore, K. Freirich, P. A. Kotos, et al., "IEEE standard inverse-time characteristic equations for overcurrent relays," IEEE Transactions on Power Delivery, vol. 14, pp. 868-872, 1999.
[8] T. Instruments, "Stellaris LM4F232 Evaluation Board User's Manual," in http://www.ti.com/lit/ug/spmu272/spmu272.pdf, 2012.

IOP Conf. Series: Materials Science and Engineering **752** (2020) 012006 doi:10.1088/1757-899X/752/1/012006

Event-triggered Adaptive Dynamic Surface Output Feedback Inverse Control for a Class of Hysteresis Nonlinear Systems

Wenbo Liu[1] Shuai Di[1] Yaoqi Dai[2] and Wenjing Cao[3]

[1]Datang Northeast Electric Power Test and Research Institute Co., Ltd. China
[2]Qingdao Hongda Textile Machinery Co., Ltd., China
[3]Changchun Central Hospital, China

1490805218@qq.com

Abstract. Aiming at the difficult problem of the nonlinear system with hysteresis, the Prandtl-Ishlinskii model is used to describe the system. A neural network-based output feedback adaptive dynamic surface inverse control (DSIC) scheme is proposed. Firstly, in order to overcome the hysteresis compensation error, an auxiliary design parameter is introduced into the observer to construct an improved high-gain k-order filter state observer to accurately estimate the unmeasurable state. Then, the inverse of the asymmetric displacement matrix of the Prandtl-Ishlinskii model is used as feedforward compensation. the unknown density function with different thresholds is not estimated, the analysis compensation error is obtained, the over-parameterization problem is effectively avoided. At the same time, the event trigger mechanism is introduced in the controller design process to discretize the continuous control signal under the condition of event triggering to improve the computational efficiency of the controller. Finally, the simulation is verified, the results shows the control scheme has higher control precision and faster control speed.

1. Introduction

In the past few decades, piezoelectric materials have been widely used in ultra-high-precision positioning systems such as micro-nano systems and metal cutting systems [1]. such as scanning probe microscopes, optical calibration, diamond lathes, active vibration control, Biologically operated equipment, etc. More attention has been given to their ability to meet ultra-high precision requirements and their versatility in a variety of applications. However, the inevitable shortcoming is the hysteresis nonlinearity present in piezoelectric material actuators, because the hysteresis is non-differentiable and multi-valued, the control system lacks compensation for hysteresis, it may exhibit undesirable characteristic. A common method of dealing with hysteresis is to establish an inverse model of hysteresis and cascade it as a compensation for the actuator before the control system; another method to use a robust adaptive scheme [2-4]. For the first method, since the hysteresis is usually unknown, it is difficult to establish an analytical error expression between the hysteresis model and its corresponding inverse model. In the second method, the unknown hysteresis is divided into linear and nonlinear parts and is generally regarded as interference processing, but the system only outputs measurable, this method is not feasible because of the nonlinear part of the hysteresis [5-6]. It may be unbounded, means there may be no upper limit to the observation error. So far, in the output feedback control system, the use of the second method is still limited.

Content from this work may be used under the terms of the Creative Commons Attribution 3.0 licence. Any further distribution of this work must maintain attribution to the author(s) and the title of the work, journal citation and DOI.
Published under licence by IOP Publishing Ltd

Considering the resource utilization and network channel limitation in the signal transmission process, an event trigger control strategy is used [6]. When the system is less interfered or the system has reached an ideal state, the controller is not correct. Control; the controller controls the system only the system deviates from the expected state. It can ensure the system has the ideal control performance, reduces network load and resource waste and improves the utilization of network channels.

2.Problem statement

Consider the following nonlinear systems with hysteresis inputs:

$$\begin{aligned} \dot{x}_i &= x_{i+1} + f_i(\bar{x}_i) + d_i(t) \\ \dot{x}_n &= b_0 w(u) + f_n(\bar{x}_n) + d_n(t) \\ y &= x_1, i = 1, \cdots, n-1, \end{aligned} \tag{1}$$

Where $\bar{x}_i := [x_1, x_2, \cdots, x_i]^T \in R^i$ are State variable, $f_i(\bar{x}_i), i = 0, 1, \cdots, n$ is unknown smooth function, $d_i(t)$ is external interference, b_0 is unknown constant, $y \in R$ is the output of the plant, $w \in R$ represents an unknown hysteresis and can be expressed in the following form,

$$w(u) = P[u](t) \tag{2}$$

Where u is the input signal, P is the hysteresis operator.

For the above nonlinear systems, we made the following assumptions

A1: Disturbance $d_i(t), i = 1, \cdots, n$ satisfy $|d_i(t)| \le \bar{d}_i$, , wherer d_i are unkonw positive parameter.

A2: Expected trajectory y_r is smooth and $y_r(0)$ can be set according to the designer, $[y_r, \dot{y}_r, \ddot{y}_r]^T$ belongs to a known compact set for all $t \ge 0$.

A3: b_0 is known parameter, without loss of generality, considered to be greater than zero.

Remark 1: A1 shows that the interference is bounded and its widely used in robust adaptive control systems [5]. A2 is commonly used in dynamic surface control. to facilitate stability analysis and to keep all signals consistent and bounded [6]. A3 is the basic need in adaptive control.

By equation (1) we have

$$\begin{aligned} \dot{x}_i &= x_{i+1} + f_i(\bar{x}_i) + d_i(t), \\ \dot{x}_n &= b_\Lambda u_d + f_n(\bar{x}_n) + b_0 d_b(t) + d_n(t), \\ y &= x_1, i = 1, \cdots, n-1 \end{aligned} \tag{3}$$

Where b_Λ is a positive parameter and satisfy $b_\Lambda = b_0 \phi'(\Lambda)$.

The RBFNN shows in (4) is used to approximate the unknown smooth function in (3),

$$f_i(\bar{x}_i) = \psi_i^T(\xi_i)\vartheta_i^* + \varepsilon_i \tag{4}$$

Where $\varepsilon_i, i = 1, \cdots, n$ is positive constant $\xi_i = \left(\hat{\hat{x}}_1, \cdots, \hat{\hat{x}}_i\right), i = 1, \cdots, n$ is approximation error of the neural network , $\hat{\hat{x}}_1, \cdots, \hat{\hat{x}}_i$ are estimated of $x_1, \cdots, x_i$, will be introduced below.

Now substitute the (4) into (3), we can get

$$\begin{aligned} \dot{x}_i &= x_{i+1} + \psi_i^T(\xi_i)\vartheta_i^* + \varepsilon_i + d_i(t), \\ \dot{x}_n &= b_\Lambda u_d + \psi_n^T(\xi_n)\vartheta_n^* + \varepsilon_n + b_0 d_b(t) + d_n(t), \\ y &= x_1, i = 1, \cdots, n-1 \end{aligned} \tag{5}$$

Converting the above form to a state space form, we have

$$\begin{aligned} &\dot{x} = Ax + \Psi^T(\xi)\vartheta^* + bu_d + B, \\ &y = e_1^T x \end{aligned} \tag{6}$$

Where $B = D_b + \varepsilon + d$, from (5) and (6) we can obtain

$$A = \begin{bmatrix} 0 & 1 & & 0 \\ 0 & & \ddots & \\ \vdots & & & 1 \\ 0 & \cdots & 0 & 0 \end{bmatrix}, d = \begin{bmatrix} d_1(t) \\ \vdots \\ d_{n-1}(t) \\ d_n(t) \end{bmatrix}, \varepsilon = \begin{bmatrix} \varepsilon_1 \\ \vdots \\ \varepsilon_{n-1} \\ \varepsilon_n \end{bmatrix},$$

$$b = \begin{bmatrix} 0 \\ \vdots \\ 0 \\ b_\Lambda \end{bmatrix}, e_1 = \begin{bmatrix} 1 \\ 0 \\ \vdots \\ 0 \end{bmatrix}, x = \begin{bmatrix} x_1 \\ x_2 \\ \vdots \\ x_n \end{bmatrix}, D_b = \begin{bmatrix} 0 \\ \vdots \\ 0 \\ d_b(t) \end{bmatrix}, \tag{7}$$

$$\vartheta^* = \begin{bmatrix} \vartheta_1^* \\ \vdots \\ \vartheta_{n-1}^* \\ \vartheta_n^* \end{bmatrix} \subset R^{\sum_{i=1}^{n} N_n}, \Psi^T(\xi) = \begin{bmatrix} \psi_1 & & \\ & \ddots & \\ & & \psi_n \end{bmatrix}$$

Where $\psi_1 = [\psi_{1,1}(\xi_1),\cdots,\psi_{1,N}(\xi_1)], \psi_n = [\psi_{n,1}(\xi_n),\cdots,\psi_{n,Nn}(\xi_n)]$ $N_i, i = 1,\cdots,n$ are the dimension of the weight vector in the RBF neural network.

Remark 2: The control objective is establish an adaptive neural network control system based on output feedback dynamic surface inversion for a class of nonlinear hysteresis systems; the output y has a good tracking effect and all signals in the control system are semi-global and ultimately uniformly bounded.

Let $A_0 = A - qe_1^T$, we rewritten (6) as

$$\begin{aligned} &\dot{x} = A_0 x + qy + \Psi^T(\xi)\vartheta^* + bu_d + B \\ &y = e_1^T x \end{aligned} \tag{8}$$

Where

$$A_0 = A - qe_1^T = \begin{bmatrix} -q_1 & 1 & & \\ -q_2 & & \ddots & \\ \vdots & & & 1 \\ -q_n & 0 & \cdots & 0 \end{bmatrix}, q = \begin{bmatrix} q_1 \\ q_2 \\ \vdots \\ q_n \end{bmatrix} \tag{9}$$

Where A_0 is Hurwitz matrix, q is a positive parameter.

From the literature [6], the following high-gain K-order filter observers are available to estimate the unmeasurable state x in equation (8)

$$\dot{v}_0 = kA_0 v_0 + \Phi^{-1} e_n u_d \tag{10}$$

$$\dot{\xi}_0 = kA_0 \xi_0 + kqy \tag{11}$$

$$\dot{\Xi} = kA_0 \Xi + \Phi^{-1}\Psi^T \tag{12}$$

Where $K \geq 1$ is a positive design parameter. e is the nth coordinate vector. Let $\Phi = diag\left\{1, k, \cdots, k^{n-1}\right\}$ and according to the equation (10) to (12) we can obtain an estimate of the following state vectors

$$\hat{x} = \Phi\xi_0 + \Phi b_\Lambda v_0 + \Phi\Xi\vartheta^* \tag{13}$$

Define the following estimation error $\epsilon = x - \hat{x}$ we have

$$\dot{\epsilon} = A\epsilon - k\Phi q\epsilon_1 + B \tag{14}$$

Where ϵ_1 is the first value of ϵ and already defined in equation (6).

Lemma 1: Considering the above high gain k-order filter state observer, the following quadratic functions are now defined

$$V_\epsilon := \epsilon^T P\epsilon \tag{15}$$

Where $P = (\Psi^{-1})^T \bar{P}\Psi^{-1}, \bar{P} = \bar{P}^T > 0$ are positive definite matrix and satisfies

$$A_0^T \bar{P} + \bar{P}A_0 = -2I \tag{16}$$

Let

$$\zeta_\epsilon := \frac{k}{\lambda_{\max}\left(\bar{P}\right)}$$
$$\delta_\epsilon := k\left(\frac{\left\|\bar{P}\right\|\left\|B\right\|_{\max}}{k^n}\right)^2 \tag{17}$$

$\| B \|_{\max}$ is the maximum value of $\| B \|$,without any $k > 1$ we have.

$$\dot{V}_\epsilon \leq -\zeta_\epsilon V_\epsilon + \delta_\epsilon \tag{18}$$

According to eqution (13) we can get the true state estimated to

$$\hat{\hat{x}} = \Phi\xi_0 + \Phi\hat{b}_\Lambda v_0 + \Phi\Xi\hat{\vartheta} \tag{19}$$

Where $\hat{b}_\Lambda$ is the estimated value of b_Λ, $\hat{\vartheta}$ is the estimated value of ϑ^*.

Remark 2: The established high-gain K-order filter observer is used to deal with bounded terms B. on the one hand, the observer error can be arbitrarily small. on the other hand by properly selecting design parameters and matrices.

3.Adaptive DSC Design

In this section the controller using an adaptive dynamic surface inverse control scheme is prososed. The control structure is shown in Figure 1.

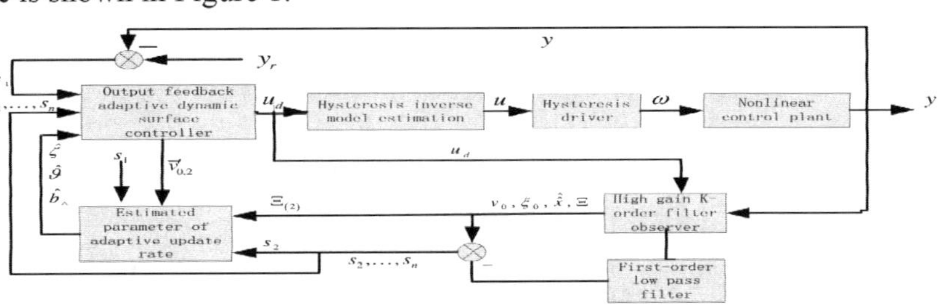

Figure 1. Control structure

Step 1: Define the first error surface S_1,

$$S_1 = y - y_r \tag{20}$$

Where y_r is expected trajectory, according (5) we can get the derivative of S_1.

$$\begin{aligned}\dot{S}_1 &= \dot{y} - \dot{y}_r \\ &= x_2 + \psi_1^T(\xi_1)\vartheta_1^* + \varepsilon_1 + d_1(t) - \dot{y}_r\end{aligned} \tag{21}$$

According to (14) ,we can get,

$$\begin{aligned}x_2 &= \hat{x}_2 + \epsilon_2 \\ &= k\xi_{0,2} + kb_\Lambda v_{0,2} + k\Xi_{(2)}\vartheta^* + \epsilon_2\end{aligned} \tag{22}$$

Where Ξ is the second vector of $\Xi_{(2)}$, substitute the (22) into (21),we can get,

$$\dot{S}_1 = k\xi_{0,2} + b_\Lambda kv_{0,2} + k\Xi_{(2)}\vartheta^* + \vartheta_1^{*^T}\psi_1(\xi_1) + \varepsilon_1 + d_1(t) - \dot{y}_r + \epsilon_2 \tag{23}$$

We rewritten the equation (23) can get,

$$\begin{aligned}\dot{S}_1 &= k\xi_{0,2} + b_\Lambda k(v_{0,2} - \overline{v}_{0,2}) + b_\Lambda k\overline{v}_{0,2} \\ &+ \vartheta^{*T}\left(k\Xi_{(2)} + \Psi_{(1)}\right)^T + \varepsilon_1 + d_1(t) - \dot{y}_r + \epsilon_2\end{aligned} \tag{24}$$

Let $\overline{v}_{0,2}$ is a virtual control signal,

$$\overline{v}_{0,2} = \hat{\zeta}\overline{v}'_{0,2} \tag{25}$$

Where $\hat{\zeta}$ is estimated of $\zeta = 1/b_\Lambda$ and we can get,

$$\overline{v}'_{0,2} = \left(-l_1S_1 - \hat{\vartheta}^T\left(k\Xi_{(2)} + \Psi_{(1)}\right)^T - k\zeta_{0,2} + \dot{y}_r\right)/k \tag{26}$$

Where l_1 is a positive design parameter, $\hat{\vartheta}$ is the estimated of ϑ^*.we choose adaptive law $\hat{\zeta}$, $\hat{\vartheta}$ as follow,

$$\begin{aligned}\dot{\hat{\zeta}} &= -\gamma_\zeta\left(k\overline{v}'_{0,2}S_1 + \sigma_\zeta\hat{\zeta}\right) \\ \dot{\hat{\vartheta}} &= \gamma_\vartheta\left[\left(k\Xi_{(2)} + \Psi_{(1)}\right)^T S_1 - \sigma_\vartheta\vartheta\right]\end{aligned} \tag{27}$$

Let $\overline{v}_{0,2}$ through the first-order low pass filter,

$$z_2 + \upsilon_2\dot{z}_2 = \overline{v}_{0,2}, z_2(0) = \overline{v}_{0,2}(0) \tag{28}$$

Where υ_2 is a positive parameter, z_2 is new state variable.

Step 2: Define the first error surface S_2

$$S_1 = y - y_r \tag{29}$$

According to (29) we can get the derivative of S_2 .

$$\begin{aligned}\dot{S}_2 &= -kq_2v_{0,1} + kv_{0,3} - \dot{z}_2 \\ &= k(v_{0,3} - \overline{v}_{0,3}) + k\overline{v}_{0,3} - kq_2v_{0,1} - \dot{z}_2\end{aligned} \tag{30}$$

Let $\overline{v}_{0,2}$ is a virtual control signal

$$\overline{v}_{0,3} = \left(-l_2S_2 + kq_2v_{0,1} + \dot{z}_2 - \hat{b}_\Lambda kS_1\right)/k \tag{31}$$

Where l_2 is a positive design parameter, we choose adaptive law $\hat{b}_\Lambda$ as follow,

$$\dot{\hat{b}}_\Lambda = \gamma_b\left(kS_1S_2 - \sigma_b\hat{b}_\Lambda\right) \tag{32}$$

Let $\overline{v}_{0,3}$ throught the first-order low pass filter

$$z_2 + \upsilon_2 \dot{z}_2 = \overline{v}_{0,2}, z_2(0) = \overline{v}_{0,2}(0) \tag{33}$$

Where υ_3 is a positive parameter, z_3 is new state variable

Step 3: $i\ (3 \le i \le n-1)$ Define the i th error surface

$$S_i = v_{0,i} - z_i \tag{34}$$

According to (34) the derivative of S_i.

$$\begin{aligned}\dot{S}_i &= -kq_i v_{0,1} + kv_{0,i+1} - \dot{z}_i \\ &= -kq_i v_{0,1} + k\left(v_{0,i+1} - \overline{v}_{0,i+1}\right) + k\overline{v}_{0,i+1} - \dot{z}_i\end{aligned} \tag{35}$$

Let $\overline{v}_{0,2}$ is a virtual control signal

$$\overline{v}_{0,i+1} = \left(-l_i S_i + kq_i v_{0,1} + \dot{z}_i\right)/k \tag{36}$$

Where l_i is a positive design parameter, we choose adaptive law $\overline{v}_{0,i+1}$ as follow

$$\dot{\hat{b}}_\Lambda = \gamma_b\left(kS_2 S_i - \sigma_b \hat{b}_\Lambda\right) \tag{37}$$

Let $\overline{v}_{0,i+1}$ throught the first-order low pass filter

$$z_2 + \upsilon_2 \dot{z}_2 = \overline{v}_{0,2}, z_2(0) = \overline{v}_{0,2}(0) \tag{38}$$

Where υ_{i+1} is a positive parameter, z_{i+1} is new state variable

Step n: Define the n th error surface

$$S_n = v_{0,n} - z_n \tag{39}$$

According to (34) we can get the derivative of S_n.

$$\dot{S}_n = -kq_n v_{0,1} + k^{1-n}\overline{v}_{0,n} - \dot{z}_n \tag{40}$$

Let $\overline{v}_{0,\mathrm{n}}$ is a virtual control signal

$$\overline{v}_{0,\mathrm{n}} = k^{n-1}\left(kq_n v_{0,1} - l_n S_n + \dot{z}_n\right) \tag{41}$$

Where l_n is a positive design parameter

Next, an event triggering mechanism will be introduced on the basis of the above control structure. First, the following adaptive controller is given:

$$\breve{u}_d(t) = \overline{v}_{0,\mathrm{n}} - \overline{g}\tanh\left(\frac{k^{1-n}S_n\overline{g}}{\rho}\right) \tag{42}$$

The trigger event:introduce as follow

$$\begin{aligned}&\breve{u}_d(t)u_d(t) = \breve{u}_d(t_j), \forall t \in [t_j, t_{j+1}) \\ &t_{j+1} = \inf\left\{t > t_j \,|\, |e(t)| \ge g\right\}, t_1 = 0\end{aligned} \tag{43}$$

Where $e(t) = \breve{u}_d(t) - u_d(t)$ is measurement error, ρ, g and $\overline{g} > g$ are positive design parameters, $t_j, j \in z^+$ is the controller update time.

Remark 3: At the moment t_{j+1}, the control signal $u_d(t)$ will be updated to $\breve{u}_d(t_{j+1})$, on the other hand ,at the moment $t \in [t_j, t_{j+1})$, the control signal remains unchanged, i.e $u_d(t) = \breve{u}_d(t_j)$.

IOP Conf. Series: Materials Science and Engineering **752** (2020) 012006 doi:10.1088/1757-899X/752/1/012006

4. Stability analysis

In this section, the stability analysis for the proposed DSC scheme will be presented, we define

$$\dot{y}_2 = \dot{z}_2 - \dot{\bar{v}}_2 = -\frac{y_2}{\upsilon_2} - \dot{\hat{\zeta}}\left(-k\xi_{0,2} - l_1 S_1 + \dot{y}_r - \hat{\vartheta}^T\left(\Psi_{(1)} + k\Xi_{(2)}\right)^T\right)/k$$
$$-\hat{\zeta}\left(-l_1\dot{S}_1 - k\dot{\xi}_{0,2} - \left[\hat{\vartheta}^T\left(\Psi_{(1)} + k\Xi_{(2)}\right)^T\right]' + \ddot{y}_r\right)/k \quad (44)$$

$$\dot{y}_{i+1} = -\frac{y_{i+1}}{\upsilon_{i+1}} + \frac{1}{k} l_i \dot{S}_i - q_i \dot{v}_{0,1} - \frac{1}{k}\ddot{z}_i, i = 3, \cdots, n-1. \quad (45)$$

Where $\bar{v}_{0,2}$ and $\bar{v}_{0,n}$ are given by (26) and (32). acording to (31) and (38), we can get

$$\dot{z}_2 = \frac{\bar{v}_{0,2} - z_2}{\upsilon_2} = \frac{\bar{v}_{0,2} - y_2 - \bar{v}_{0,2}}{\upsilon_2} = -\frac{y_2}{\upsilon_2} \quad (46)$$

Therefore, the derivative of (45) can be written as

$$\dot{y}_2 = \dot{z}_2 - \dot{\bar{v}}_2 = -\frac{y_2}{\upsilon_2} - \dot{\hat{\zeta}}\left(-k\xi_{0,2} - l_1 S_1 + \dot{y}_r - \hat{\vartheta}^T\left(\Psi_{(1)} + k\Xi_{(2)}\right)^T\right)/k$$
$$-\hat{\zeta}\left(-l_1\dot{S}_1 - k\dot{\xi}_{0,2} - \left[\hat{\vartheta}^T\left(\Psi_{(1)} + k\Xi_{(2)}\right)^T\right]' + \ddot{y}_r\right)/k \quad (47)$$

Now define the following Lyapunov function:

$$V = \frac{1}{2}\sum_{i=1}^{n} S_i^2 + \frac{1}{2}\sum_{i=1}^{n-1} y_{i+1}^2 + \frac{b_\Lambda}{2\gamma_\zeta}\tilde{\zeta}^2 + \frac{1}{2\gamma_\vartheta}\tilde{\vartheta}^T\tilde{\vartheta} + \frac{1}{2\gamma_b}\tilde{b}_\Lambda^2 + V_\epsilon \quad (48)$$

We can get the derivative of V as follow

$$\dot{V} \le -\left(l_1 - \frac{2 + b_\Lambda k}{2} - \frac{k^2}{2}\right)S_1^2 - (l_2 - k)S_2^2 + \sum_{i=3}^{n-1}\left[-\left(l_i - \frac{3k}{2}\right)S_i^2\right]$$
$$-\left(l_n - \frac{k}{2}\right)S_n^2 - \frac{\sigma_\vartheta}{2}\tilde{\vartheta}^T\tilde{\vartheta} + \frac{\sigma_\vartheta}{2}\vartheta^{*T}\vartheta^* - \frac{b_\Lambda\sigma_\zeta}{2}\tilde{\zeta}^2 + \frac{b_\Lambda\sigma_\zeta}{2}\zeta^2$$
$$-\frac{\sigma_b}{2}\tilde{b}_\Lambda^2 + \frac{\sigma_b}{2}b_\Lambda^2 - \left(\frac{1}{\upsilon_2} - \frac{M_2^2}{2\kappa} - \frac{b_\Lambda^2 k^2}{4}\right)y_2^2 \quad (49)$$
$$-\sum_{i=2}^{n-1}\left[\left(\frac{1}{\upsilon_{i+1}} - \frac{M_{i+1}^2}{2\kappa} - \frac{k}{2}\right)y_{i+1}^2\right] - \left(\frac{k}{\lambda_{\max}(\bar{P})} - \frac{1}{2\lambda_{\min}(\bar{P})}\right)V_\epsilon$$

Choose the following design parameters below

$$
\begin{aligned}
& k \geq \lambda_{\max}\left(\bar{P}\right)C_1 + \frac{\lambda_{\max}\left(\bar{P}\right)}{2\lambda_{\min}\left(\bar{P}\right)}, \\
& l_1 \geq \frac{2+b_\Lambda k}{2} + \frac{k^2}{2} + C_1, l_2 \geq k + C_1, \\
& l_i \geq \frac{3k}{2} + C_1, i = 2,\cdots,n-1, \\
& l_n \geq \frac{k}{2} + C_1, \frac{1}{\upsilon_2} \geq \frac{b_\Lambda^2 k^2}{4} + \frac{M_2^2}{2\kappa} + C_1, \\
& \frac{1}{\upsilon_{i+1}} \geq \frac{k}{2} + \frac{M_{i+1}^2}{2\kappa} + C_1, i = 2,\cdots,n-1,
\end{aligned} \tag{50}
$$

Substituting equations (48) and (50) into (49), we have

$$\dot{V} \leq -2C_1 V + C_2 \tag{51}$$

Where $C_2 = \left(n-1\right)\kappa/2 + \sigma_\vartheta \vartheta^{*T}\vartheta^*/2 + b_\Lambda \sigma_\varsigma \zeta^2/2 + \sigma_b b_\Lambda^2/2 + \left(\varepsilon_1^2 + d_1^2 + 2\delta_\epsilon\right)/2 + 0.2785\rho$ is a positive design parameter, and satisfy $C_1 \geq C_2/2p$

Let $S_1(0)=0$, $S_i\left(0\right)=0, i=3,\cdots,n$, $y_i\left(0\right)=0, i=2,\cdots,n$, we can get the upper boundary $V(t)$

$$V\left(t\right) \leq \frac{C_2}{2C_1} + \frac{1}{k^2}\lambda_{\max}\left(\bar{P}\right)\left\|\epsilon\left(0\right)\right\|^2 \tag{52}$$

It can be further seen from the above equation, the L_∞ norm tracking error is shown in (52).

$$
\begin{aligned}
\left\|S_1\right\|_\infty &= \sup_{t\geq 0}\left|S_1\right| = \left\|x_1 - y_r\right\|_\infty \leq \left\|\sqrt{2V}\right\|_\infty \\
&\leq \sqrt{\frac{C_2}{C_1} + \frac{2}{k^2}\lambda_{\max}\left(\bar{P}\right)\left\|\epsilon\left(0\right)\right\|^2}
\end{aligned} \tag{53}
$$

According to (53), the appropriate selection of design parameters $\| S_1 \|_\infty$ can be arbitrarily small. It is now proven the event triggering mechanism, exist $t^* > 0$ to make time interval $\{t_{j+1} - t_j\} > t^*, \forall j \in z^+$ have a lower bound, and according to (42) we can get

$$\dot{\breve{u}}_d = \dot{\bar{v}}_{0,n} - \frac{\bar{g}^2 \dot{S}_n}{\rho\cosh^2\left(\frac{k^{1-n} S_n \bar{g}}{\rho}\right)} \tag{54}$$

From (54) we know $\dot{\breve{u}}_d$ is a continuous function, so $\left|\dot{\breve{u}}_d\right|$ is bounded.

$$\dot{\breve{u}}_d = \dot{\bar{v}}_{0,n} - \frac{\bar{g}^2 \dot{S}_n}{\rho\cosh^2\left(\frac{k^{1-n} S_n \bar{g}}{\rho}\right)} \tag{55}$$

5.Simulation Results

This section will apply a second-order hysteresis nonlinear system to verify the effectiveness of the above control method.

Consider the following nonlinear system with an unknown hysteresis input

IOP Conf. Series: Materials Science and Engineering **752** (2020) 012006 doi:10.1088/1757-899X/752/1/012006

$$\begin{aligned}&\dot{x}_1 = x_2 + 0.4\sin(x_1)x_1^2 + 0.3x_1^2 + 0.1\cos(t)\\ &\dot{x}_2 = w + 0.8\cos(x_1)x_2 + x_1\sin(x_2) + 0.3\sin(t)\\ &y = x_1\end{aligned} \tag{56}$$

Where w representative hysteresis nonlinearity. The corresponding density function is selected as $p(r) = e^{-0.067(r-1)^2}, r \in [0,5]$. The goal of this simulation is to make the system output a better tracking reference signal $y_r = 0.5\sin(0.5\mathrm{t}) + 0.5\cos(\mathrm{t})$.

The design parameters of the state observer are selected as follows:

$$\begin{aligned}&q = \begin{bmatrix} q_1 \\ q_2 \end{bmatrix} = \begin{bmatrix} 5 \\ 3 \end{bmatrix},\\ &A_0 = \begin{bmatrix} -q_1 & 1 \\ -q_2 & 0 \end{bmatrix}, \Phi^{-1} = \begin{bmatrix} 1 & 0 \\ 0 & 1/k \end{bmatrix}\end{aligned} \tag{57}$$

In the simulation, the initial value of the design parameter value of the state and the adaptive rate is selected as $x_1(0) = x_2(0) = 0, \hat{\zeta}(0) = 0, \hat{\vartheta}(0) = 0, \hat{b}_\Lambda(0) = 0.1$. Radial basis function neural network $\psi_1(\xi_1), \psi_2(\xi_2)$,each of the 7 and 11 nodes is selected, and the centers of the basis functions are respectively $-1, -0.8, -0.4, 0, 0.4, 0.8, 1$;The corresponding width is $\eta_i = 1, i = 1, \cdots, 7$; $\xi_1 = (\hat{\hat{x}}_1)$ and $\xi_2 = (\hat{\hat{x}}_1, \hat{\hat{x}}_2)$,In addition,

$$\begin{aligned}\Psi^T(\xi) &= \begin{bmatrix} \psi_1 & 0 \\ 0 & \psi_2 \end{bmatrix}\\ &= \begin{bmatrix} \psi_{1,1}(\xi_1), \cdots, \psi_{1,7}(\xi_1) & 0 \\ 0 & \psi_{2,1}(\xi_1), \cdots, \psi_{2,7}(\xi_2) \end{bmatrix}\end{aligned} \tag{58}$$

The simulation results shown in figure 2-4.

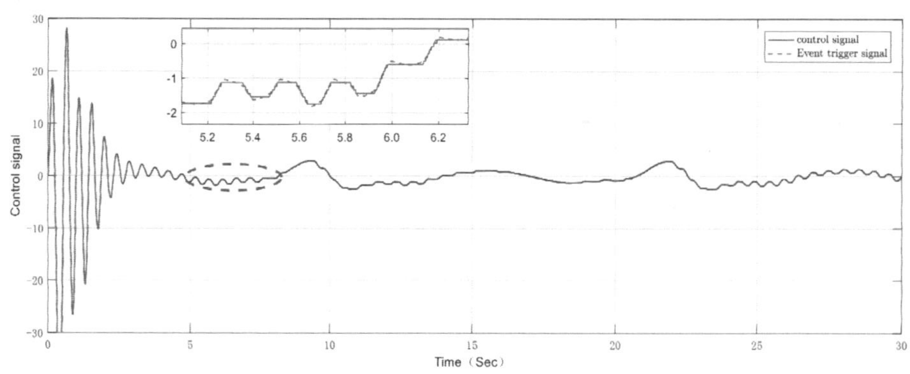

Figure 2. Control signal

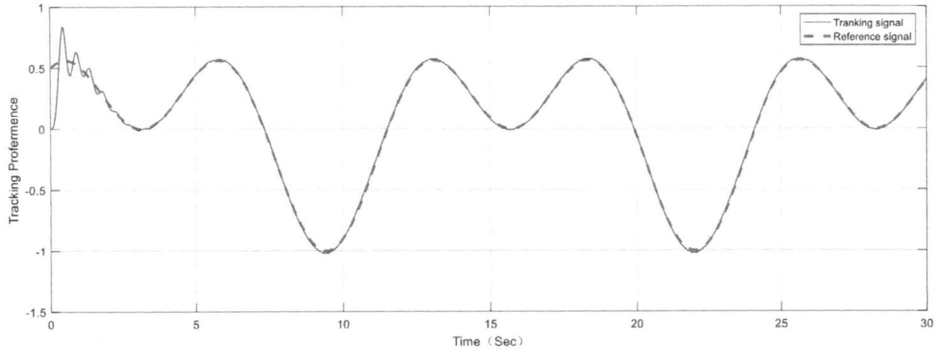

Figure 3. Tracking performance

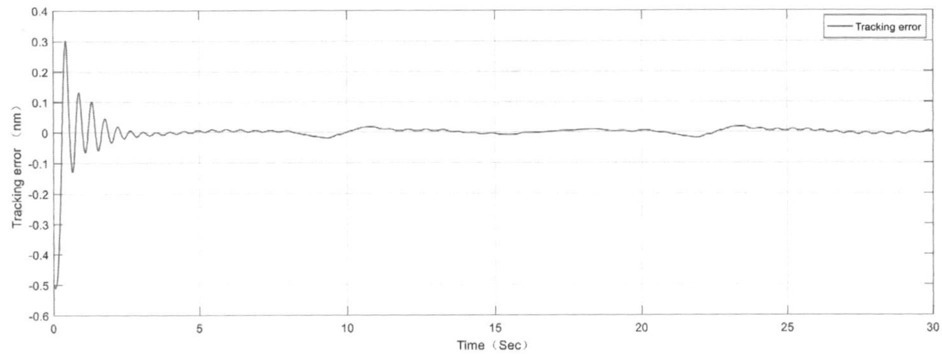

Figure 4. Tracking error

6.Conclusion

The dynamic surface method can avoid the "differential explosion" problem caused by the backstepping method and reduce the computational burden. The stability analysis proves that all signals of the closed-loop system are semi-global and ultimately uniformly bounded.

Reference

[1] Helmers M, Herrmann M. 2018 Hysteresis and Phase Transitions in a Lattice Regularization of an Ill-Posed Forward–Backward Diffusion Equation Archive for Rational Mechanics & Analysis, vol 230 pp 231-275.
[2] Safadi M M, Rubin M B. Modeling rate-independent hysteresis in large deformations of preconditioned soft tissues International Journal of Solids & Structures, 2014, vol 51 pp 3265-3272.
[3] Dimian M, Andrei P 2003 Mathematical Models of Hysteresis Signals & Communication Technology, 2003, vol 22 pp 603-608.
[4] Sato M. Mechanical properties of living tissues. 1986 Iyō Denshi to Seitai Kōgaku Japanese Journal of Medical Electronics & Biological Engineering vol 24 pp 213-219.
[5] Li Y, Tong S 2017 Adaptive Fuzzy Control with Prescribed Performance for Block-Triangular-Structured Nonlinear Systems IEEE Transactions on Fuzzy Systems, pp 1-1.
[6] Heemels W P M H, Johansson K H, Tabuada P. An Introduction to Event-triggered and Self-triggered Control. Conference on Decision and Control. 2009. P 3

IOP Conf. Series: Materials Science and Engineering **752** (2020) 012007 doi:10.1088/1757-899X/752/1/012007

Renewable Energy Matching to Consumer Entities Based on a Maximum Flow Model

Junghoon Lee[a] and Gyung-Leen Park[b]

Dept. of Computer Science and Statistics, Jeju National University, Republic of Korea

[a]jhlee@jejunu.ac.kr; [b]glpark@jejunu.ac.kr

Abstract. This paper designs a renewable energy assignment scheme based on maximum flow modeling for a microgrid containing distributed energy sources and consumption entities. Between the source and the sink node of a directed graph, mainly solar generators and consumption entities are interconnected via weighted links. Weights from the source to generators are decided by the amount of solar power production while weights from consumption entities to the sink by the amount of their energy demands. In addition, weights from generators to consumers correspond to the transmission line capacity connecting two objects, each from both parties. An additional node is defined to present the energy flow contracted with the main grid and its weight can be adjusted according to the additional energy purchase. The performance measurement result obtained from a prototype implementation shows that the renewable energy is stably distributed over multiple consumers until the energy generation reaches the maximum link capacity.

1. Introduction

In smart cities, many smart services are provided to the citizens, taking advantage of advanced information technologies and automated control mechanisms [1]. One of their most important targets is the smart power system, or smart grid, which pursues citywide energy efficiency [2]. Efficient algorithms, communication channels between diverse objects, sensor devices provide indispensable building blocks for smart grids [3]. In the modern power systems, the integration of renewable energies such as sunlight in addition to the appearance of complex power load such as EVs (Electric Vehicles) makes it necessary to establish a more sophisticated control strategy [4]. Considering the unavoidable intermittent nature of renewable energy sources, it is desirable to consume as much energy as possible within an autonomous power subsystem called *microgrid* [5]. It is quite complicated and expensive to tightly control the flow of individual overproduced energy to the main grid.

Basically, microgrids are connected to the main grid to cope with the energy lack which takes place when the demand exceeds the embedded power generation capacity. In addition, a microgrid has a sort of renewable energy generators and energy storage equipment, trying to locally control the demand and supply with just minimized intervention of the main grid [6]. Renewable energy generators can send energy only to those consumers having direct cable connections. Here, every generator and consumption entity is not connected to the backbone power transmission network due to geographic constraints or economic reasons. That is, some consumer devices depend only on the renewable energy, risking the suspension of their operations. After all, the microgrid power network leads to a network flow graph, which allows us to develop an efficient energy allocation mechanism for the

Content from this work may be used under the terms of the Creative Commons Attribution 3.0 licence. Any further distribution of this work must maintain attribution to the author(s) and the title of the work, journal citation and DOI.
Published under licence by IOP Publishing Ltd

IOP Conf. Series: Materials Science and Engineering **752** (2020) 012007 doi:10.1088/1757-899X/752/1/012007

microgrid. By this, it possible to maximize the amount of renewable energy consumed locally as well as to minimize the energy exchange with the main grid.

2. Renewable energy allocation

2.1. Network flow modeling

The energy consumption model can be represented by a directed graph from a source to a sink, while each generator and consumer entity is mapped to a node as shown in Figure 1. In the network flow graph, only those consumer entities receiving energy from at least one of renewable sources are included. It doesn't matter whether they get energy also from the microgrid backbone. To begin with, there are links from the source to all respective generator nodes and also from all respective consumer entity nodes to the sink. Excluding the source and sink nodes, the graph consists of two parties, one for generators in the left column and the other for consumers in the right column. Links come from nodes in the generator group to nodes in the consumer group, according to the existence of transmission cable connection.

The energy demand of a consumer entity, such as *A*, *B*, or *C* in Figure 1, is denoted by the weight of its link to the sink node. For a consumer entity, the sum of the weight for each input link must be larger than the weight of outgoing link to be able to meet its energy demand. Here, the electricity system is stable when the amount of actual flow is equal to the link weight which corresponds to the energy demand, for all consumer entity nodes. Exceptionally, node *A* in Figure 1 takes energy only from a renewable energy generator, so it can fail in meeting the energy demand. The input link capacity can be less than the actual demand irrespective of the installed electricity transmission potential in case the energy generation condition is not good. In that case, *A* will highly likely start discharging its battery, enter the energy-saving mode, or even stop its operation. On the contrary, *B* and *C* get energy from both the backbone and distributed sources, just trying to take from the latter as much as possible.

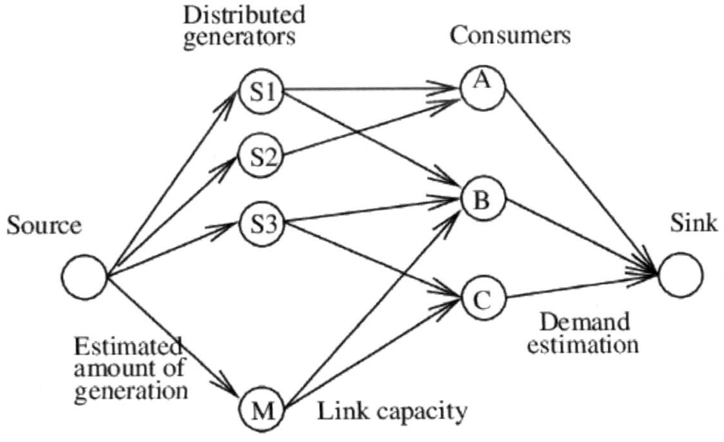

Figure 1. Network flow model.

The link weight from the source to a generator node, marked S_1, S_2, and S_3 in Figure 1, is the amount of electricity created in the generator, while the value is constrained by its maximum capacity. The generator capacity depends on the solar panels, battery chargers, and the like. The commonly available solar energy facility can produce tens or hundreds of *kw*. Particularly, *M* is the node representing the energy flow from the backbone power system, while the link capacity from the source to *M* is the amount of provisioned energy in the microgrid. Its electricity comes from the main grid or stable sources such as thermal power plants. The link weight from a power generator to a consumption entity is equal to the transmission cable capacity installed between both objects. Generated electricity can be distributed over the multiple consumers. On the contrary, a single consumer can receive from multiple heterogeneous generators.

Based on this model, it is possible to make an energy supply plan by deciding the weight of each link [7]. The link weight is fixed for those links from solar energy generators to each consumption entity node, as they are hardware-dependent maximum amount of flow. The link weight from the source to a solar generator is the predicted value of power generation and the prediction much depends on the weather condition. In addition, the energy demand is also forecasted on each consumption entity based on the current consumption level and other factors just like in the frequency control mechanism. The link weight from the source to *M* is basically the amount of maximum energy flow contracted between the main grid and the microgrid. Usually, the link weight from *M* to a consumption node must be large enough to meet the consumer demand even in the absence of available renewable energy. For those links from the source to generator nodes, their link weights are limited by the generator capacity and the flow cannot exceed this limit.

The power demand in the consumer side keeps changing, however, the energy storage equipment can possibly absorb demand spikes, making the load flat for the most time interval and thus we can balance demand and supply based on the average demand level. Figure 2 shows an example in which S_1 generates 5 units of energy, while it can send to *A* up to 5 and to *B* up to 4. However, due to the actual demand and the allocation strategy, 2 units flow to *A* and *B*, respectively. 1 unit, not consumed in S_1 can be discarded or sold to the main grid. The actual energy flow in Figure 2(b) cannot exceed the link capacity marked in Figure 2(a).

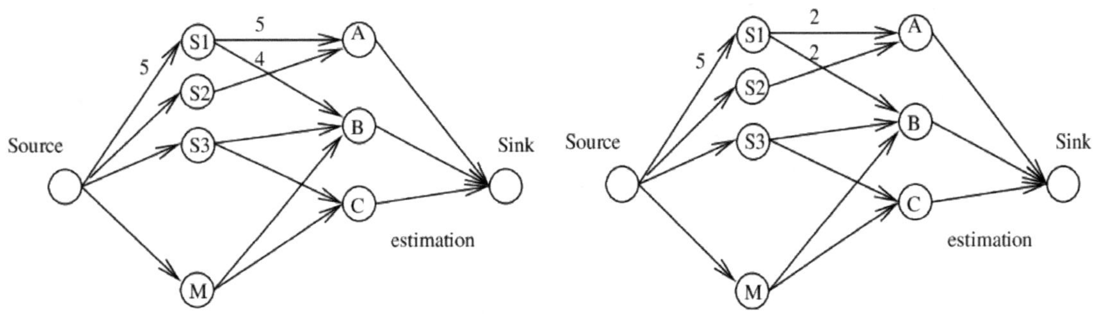

(a) Link capacity (b) The amount of flow

Figure 2. Actual flow description.

2.2. 2-pass assignment

Once the network flow model is built, particularly with the weight of each link, we can calculate the maximum amount of renewable energy for the microgrid to consume, taking advantage of the Ford-Fulkerson algorithm [8]. Then, it is also possible to know how much electricity should be assigned to respective consumers from each generator node. This scheme iterates the breadth or depth first search until there is no further path from source to sink. In each iteration finding a new path, the amount of feasible energy flow for a path is the minimum of the weights along the path. The weights of links on the path are decreased by this minimum value. Here, the process also increases the weight of every reverse link to allow potential selection canceling in finding the maximum flow, even though the reverse links are not explicitly shown in the network flow graph in Figure 1.

After all, we can know the total amount of energy flow and if this amount is smaller than the total demand in the microgrid (except for the demand which depends only on renewable sources), it is necessary to compensate for the insufficiency by purchasing additional energy from the main grid. How much to purchase more can be calculated by summing up the remaining weights in outlinks of consumer nodes. Now, it is necessary to increase the weight of the link from the source to *M*. Next, the Ford-Fulkerson is run again to meet the energy demand and allocate the needed energy to each consumption entity. The remaining outlink weight for *M* becomes 0. After the secondary execution of the network flow process, it is possible that the link weight from a solar energy generator may still

IOP Conf. Series: Materials Science and Engineering **752** (2020) 012007 doi:10.1088/1757-899X/752/1/012007

remain when it cannot be taken by any consumers. The amount of such energy is the sum of remaining weights in outlinks of generator nodes. Such electricity can be stored in the battery device (possibly including electric vehicles) or sold back to the main grid if connected [9].

3. Performance analysis

This section measures the performance of the proposed scheme via prototype implementation with Java programming. We select a microgrid having 5 solar energy generators and 10 consumption entities receiving energy from renewable sources. Without loss of generality, each consumption entity demands the same units of energy, namely, 20 units. Here, the unit is not explicitly specified, as it can be different by the energy consumption level of the facility. Each generator has connections to 3 out of 10 consumers, while the transmission capacity for all generator-to-consumer links is fixed to 10 units. In addition, 7 out of 10 consumers take energy also from the main grid with the link capacity of 20 units to cover the whole demand even without the renewable energy. The basic contract between the main grid and the target microgrid is set to 50 units. With this environment setting, we measure the effect of the amount of renewable energy produced in the generators.

The first experiment measures the amount of energy to additionally purchase from the main grid according to the increase in the per-generator solar energy creation. As almost all electricity is consumed in the microgrid, the energy purchase linearly decreases until 23 units as shown in Figure 3. Then, the energy is overproduced and there is no need to purchase any more. This result indicates that the network flow model can appropriately distribute electricity to respective consumers, calculating the amount of energy to purchase quite predictably. The maximum link weight from the source to *M* is assumed to be large enough to accommodate the demand in the microgrid. Otherwise, in case no renewable energy can be produced, the microgrid may undergo blackout.

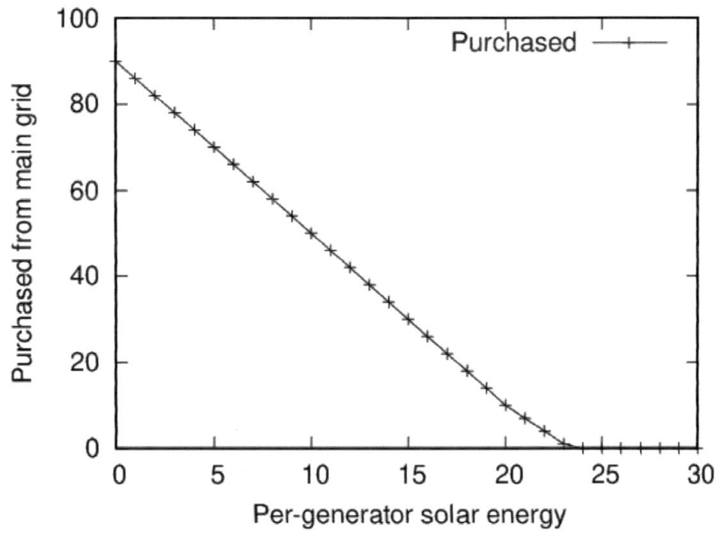

Figure 3. Purchased electricity

Next experiment measures the unmet demand for those nodes taking energy only from solar generators. Actually, even if sufficient electricity is created, the demand cannot be met in case the link weight from generators is less than the demand. Figure 4 shows this situation. Even if the amount of renewable energy generation goes beyond 25 units, consumers suffer from energy insufficiency. Until then, the insufficiency decreases according to the increase in the generation. The sharp drop in the curve after 30 units indicates that more energy is assigned to the nodes solely connected to renewable sources.

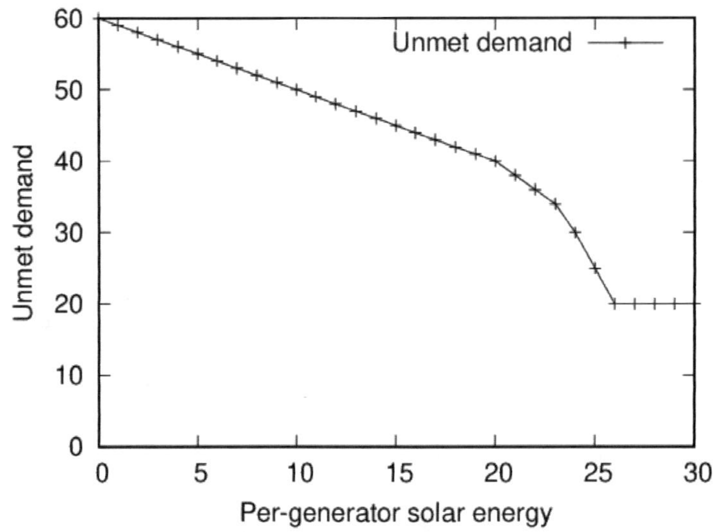

Figure 4. Unmet demand

4. Conclusions

In this paper, we have designed a renewable energy assignment scheme for a microgrid. Here, the network flow graph model is built to incorporate transmission cable connection, interaction with the main grid, distributed energy sources, and consumer demand. The two-pass maximum flow scheme first calculates the insufficiency in the microgrid and then the amount of surplus energy remaining after compensating for the lack. Here, the Ford-Fulkerson algorithm is exploited to obtain the maximum flow from renewable energy generators to respective consumer devices. Integration of intelligent computer algorithms in the smart city enriches smart services in our daily lives. As future work, we are planning to develop an elaborate solar energy forecast model to integrate into the network flow graph. This begins with the collection of weather information archive, solar energy generation history, and facility details including geographic records.

Acknowledgments

Following are results of a study on the "Leaders INdustry-university Cooperation+" Project, supported by the Ministry of Education, (MOE), Republic of Korea.

References

[1] Colistra, J.: The evolving architecture of smart cities. In: Proc. 4th IEEE International Smart Cities Conferences (2018)

[2] Masera, M., Bompard, E., Profumo, F., and Hadjsaid, N.: Smart (electricity) grid for smart cities : Assessing roles and societal impacts. Proceeding of the IEEE, Vol. 106, No. 4, 613-625 (2016)

[3] Cintuglu, M., Mohammed, O., Akkaya, K., and Uluagac, A.: A survey on smart grid cyber-physical system testbeds. IEEE Communications Surveys & Tutorials, Vol. 19, Issue 1, 446-464 (2017)

[4] Morstyn, T., Teytelboym, A., McCulloh, M.: Bilateral contract networks for peer-to-peer energy trading. IEEE Transactions on Smart Grid, Vol. 10, No, 2, 2026-2035 (2019)

[5] Shi, Z., Liang H., Huang, S., Dinavahi, V.: Distributionally robust chance-constrained energy management for islanded microgrids. IEEE Transactions on Smart Grid, Vol. 10, No, 2, 2234-2244 (2019)

[6] Rahbar, L., Chai, C., Zhang, R.: Energy cooperation optimization in microgrids with renewable energy integration. IEEE Transactions on Smart Grid, Vol. 9, No, 2, 1482-1493 (2018)

[7] Benda, D., Chu, X., Sun, S., Quek, T., and Buchley, A.: Renewable energy sharing among base stations as a min-cost-max-flow optimization problem. IEEE Transactions on Green Communication and Networking. Vol. 3, No.1, 67-78 (2019)
[8] Cormen, T., Leiserson, C., Rivest, R., and Stein, C.: Introduction to Algorithms, MIT Press and McGraw-Hill (2001)
[9] Lee J. and Park, G.: Charger reservation scheme for electric vehicles based on stable marriage problem adaptation. In: International Conference on Information Technology: IoT and Smart City, pp.316-319 (2018)

IOP Conf. Series: Materials Science and Engineering **752** (2020) 012008 doi:10.1088/1757-899X/752/1/012008

Study on Control Strategy of Microbus Power System Based on Branch-and-Bound Algorithm

Liwen Chong[1], Chunjian Kang[1,*], Zhang Zhang[1] and Jianwen Huang[1]

[1]National Computer Network Emergency Response Technical Team /Coordination Center of China, Beijing, China

[*]Corresponding author: vithon@163.com

Abstract. With the development of information technology, the scale of data center infrastructure has become larger and larger, which has always brought many challenges, such as unbalanced power load, high energy consumption, small scalable space and so on. In recent years, a distributed power supply structure, namely microbus power supply, has emerged, which is different from the traditional power distribution mode in computer room. It can adjust the power load distribution online in real time. This paper analyses the advantages of this micro bus structure power supply mode, and optimizes the cabinet power collection strategy based on branch and bound algorithm. By modeling and simulating the power consumption of the computer room in a real data center, the experimental results show that the proposed algorithm can greatly reduce the unbalance rate of the three-phase power supply equipment in the upper end of the computer room. Compared with the traditional power supply mode, it can reduce the maximum power imbalance by 49.88%.

1. Introduction

With the continuous development of communication services and Internet applications, the construction of IDC (Internet Data Center) has become one of the key business of communication enterprises to meet the business needs of accelerated development in the next few years. With the continuous expansion of business and scale, the input and output of data center or computer room infrastructure plays an important role in the business development of communication enterprises. However, the high energy consumption of communication industry has become a hot topic. The harmonious development of data center and environment, society and economy will face a severe test. On June 28, 2016, the State Administration of Organizational Affairs and the Development and Reform Commission compiled the 13th Five-Year Plan for Energy and Resources Conservation in Public Institutions [1]. In the plan, we need to strengthen the management of energy saving in the computer room. We must build a computerized monitoring and control system for the energy consumption and environment of the computer room, improve the energy saving management level of the data center, carry out the pilot project of the green data center, implement the energy saving transformation of the data center, and require that the average energy consumption of the data center be reduced by 8% after the transformation.

This paper focuses on the analysis and research of the rational distribution and utilization of power resources in computer room. Energy saving can be achieved by improving the utilization rate of electric energy. The main contributions are as follows: the analysis of the new power distribution mode of microbus [2] which someone else call it the track busway in computer room, and the optimization

Content from this work may be used under the terms of the Creative Commons Attribution 3.0 licence. Any further distribution of this work must maintain attribution to the author(s) and the title of the work, journal citation and DOI.
Published under licence by IOP Publishing Ltd

of power distribution based on branch and bound method, aiming at improving the utilization efficiency of power in computer room to the greatest extent on the basis of ensuring the balance of three-phase power supply.

The rest of this paper is organized as follows: Section 2 introduce the backgrounds which analyze the disadvantages of the traditional power system. In Section 3, we introduce the new type of power supply mode that is microbus in computer room. In Section 4, we model the microbus power distribution problem as an integer programing problem, and propose a branch-and-bound based algorithm to optimize the power supply within microbus. In Section 5, we present the experimental results. Finally, conclusions are provided in Section 6.

2. Backgrounds

At present, China's data centers are becoming larger and larger, more and more equipment, more and more complex systems, so the demand for power supporting facilities, especially power system, is also increasing. The traditional power supply mode of cabinet in data center is mainly composed of Uninterruptible Power Supply, header cabinet, PDM (Power Distribution Module), PDU (Power Distribution Unit), etc. Through UPS power supply to the header cabinet, and then from the header cabinet to the rack. It can hardly be modified after the construction, and the rated capacity and potential of the cabinet must be confirmed before construction, resulting in inconvenience for later use. The above-mentioned traditional power supply mode not only easily leads to the three-phase imbalance of power supply, but also restricts the flexible use of racks. Moreover, when the room is expanded and transformed, the traditional power supply system has almost no possibility to benefit the old. The traditional power supply mode will face enormous challenges. The following main problems exist in the traditional power supply mode.

2.1 Electricity consumption of cabinet cannot be reasonably allocated

Because there are many kinds of information system equipment, and the power consumption and volume cannot be unified, some equipment is small but the power consumption is large, some equipment is large and the power consumption is small. After installation, the power consumption of cabinet will vary greatly. There will often be a phenomenon that rack space is full but the power consumption is low, and the power consumption is full but the rack interior space is rich, which leads to the power and space waste.

2.2 Three-phase imbalance

Three-phase imbalance refers to the large deviation of current or voltage amplitude of three-phase electricity (phase A, phase B and phase C) in power system. If the amplitude of current or voltage of three-phase current is the same and the difference of phase angle is $2\pi/3$, it is called three-phase equilibrium or three-phase symmetry. In the three-phase imbalanced state of power supply system, there are a large number of zero-sequence [3] or negative-sequence [4] components in the system, which not only cause energy damage, but also bring a lot of adverse effects on equipment in the whole power system, which may endanger the security of the whole power system, and have some specific hazards including but not limited to increasing line loss, reducing the service life of transformers as well as the uninterruptible power supply (UPS) [5] equipment, and Influencing the operation safety of IT equipment (such as large servers, storage servers, etc.) in computer room.

3. Analysis of Power Supply Mode of New Type Microbus in Computer Room

In order to solve the above problems, in recent years, a microbus power supply mode which someone else call it the track busway has emerged in the computer room, which provides a feasibility for ensuring three-phase balance of power supply, improving the efficiency of cabinet utilization and reducing investment costs. This kind of power supply adopts the principle of distributed system design, which can distribute the equipment in the head cabinet. As shown in figure 1:

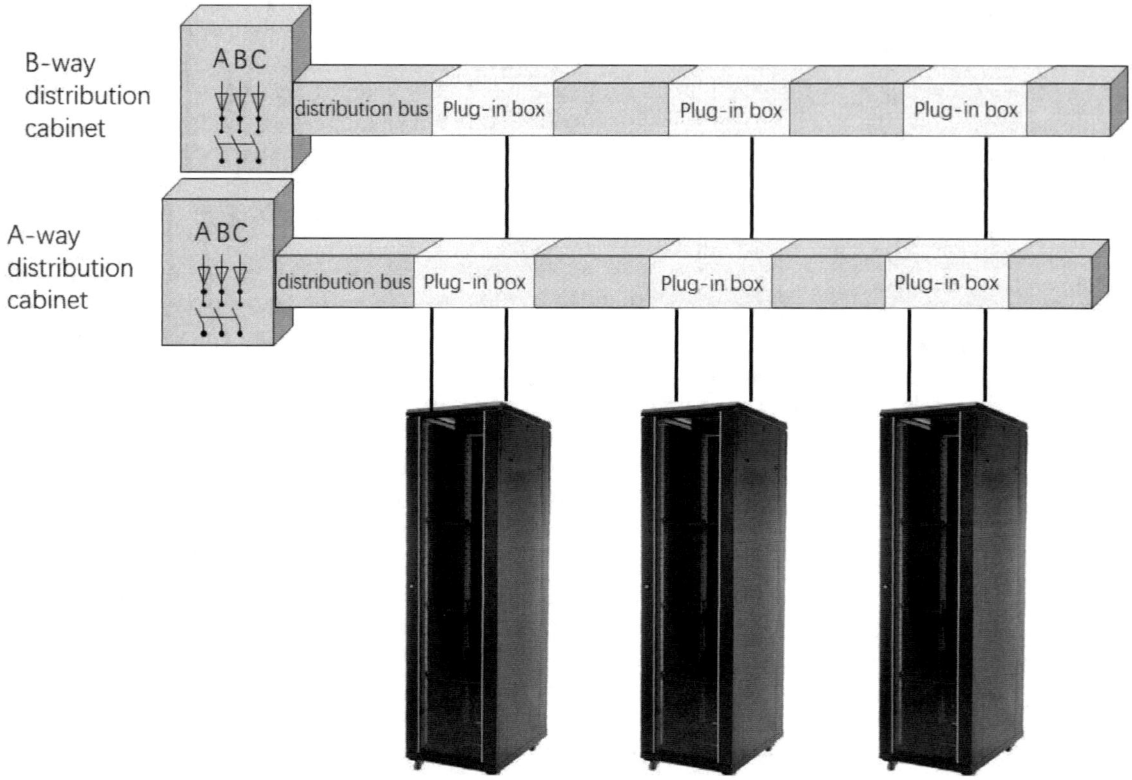

Figure 1. Schematic diagram of distributed microbus power supply system for racks.

From figure 1, it can be seen that all parts of the header cabinet are distributed installed, including overhead distribution cabinet (incoming cabinet), distribution bus, plug-in box (or terminal), branch switches and power supply detection system.

By using busbar to supply power to the racks, it is equivalent to draining water from a water tank, so it can control the reasonable output of power well. The main advantages are as follows:

(1) When the power load of one rack is lower than the rated power, the other racks under the same-phase power supply in this column can use the power beyond the rated power, as long as the total current of this phase does not exceed the rated current of the power bus.

(2) The racks supplied by different phases can also be flexibly adjusted and distributed. Because of the use of buses for power supply, the connection mode of branch cables is more flexible, and can be connected by plug-in boxes or industrial connectors. Therefore, when the power supply of one phase deviates from other phases, the balance of three-phase power supply can be guaranteed by commutating the rack.

(3) For the rack powered by single machine and double bus, the above-mentioned method can also be used to ensure load balance between two or more UPS.

4. Microbus modeling and Power Distribution Strategy Based on Branch and Bound Algorithm

In the data center, a set of UPSs will supply power for multiple racks. In the UPS output cabinet, a three-phase output switch is usually used for only one column of racks. The racks only use single-phase electricity, the number of racks per column is generally three times the number of integers, such as 9, 12, etc., so as to ensure that each column of racks can achieve the optimization goal. Bus-bar power supply system is as figure 2.

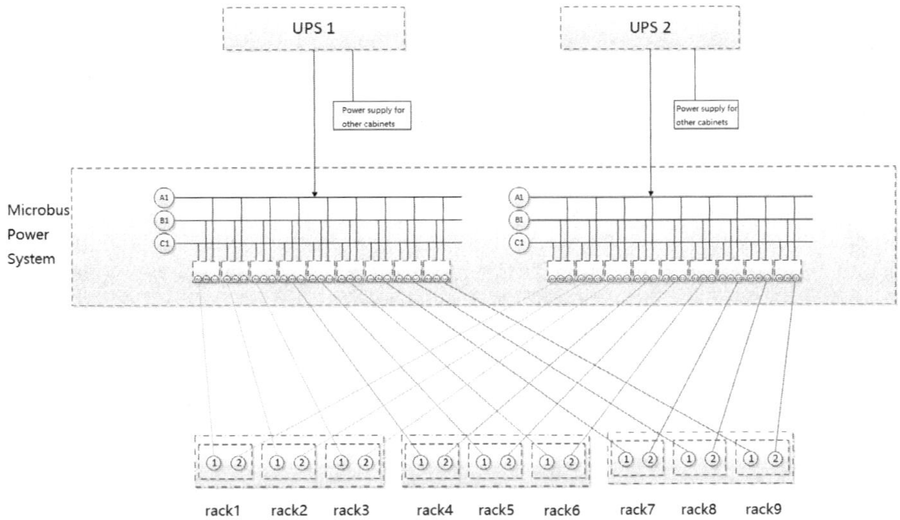

Figure 2. Microbus power supply system.

4.1 Modeling of The Power Distribution in Microbus

The optimization objective of this paper is to ensure load balance between two UPS with single machine and double bus, and load balance among different phases of UPS. The equation is as follows:

$$s = min(w_1|Q_1 - Q_2| + w_2(D_1 + D_2)) \quad (1)$$

$$Q_i = \sum_{j=1}^{3}\sum_{r=1}^{n_{row}}\sum_{k=1}^{n_r}\sum_{p=1}^{2} x_{ijrkp} \bullet load_{rkp}, i = 1,2; \quad (2)$$

$$Q_{ij} = \sum_{r=1}^{n_{row}}\sum_{k=1}^{n_r}\sum_{p=1}^{2} x_{ijrkp} \bullet load_{rkp}, i = 1,2, j = 1,2,3; \quad (3)$$

$$D_i = \frac{1}{3}\sum_{j=1}^{3}(Q_{ij} - \frac{1}{3}\sum_{j=1}^{3} Q_{ij})^2, i = 1,2; \quad (4)$$

$s.t.$

$$x_{ijrkp} \in (0,1); \quad (5)$$

$$\begin{cases} \sum_{j=1}^{3}\sum_{p=1}^{2} x_{ijrkp} = 1, i = 1,2, r = 1,\ldots,n_{row}, k = 1,\ldots,n_r; \\ \sum_{i=1}^{2}\sum_{j=1}^{3} x_{ijrkp} = 1, r = 1,\ldots,n_{row}, k = 1,\ldots,n_r, p = 1,2; \end{cases} \quad (6)$$

$$\sum_{k=1}^{n_r}\sum_{p=1}^{2} x_{ijrkp} \bullet load_{rkp} \le C_{ijr}, i = 1,2, j = 1,\ldots,n_{ps}, r = 1,\ldots,n_{row}; \quad (7)$$

$$\sum_{r=1}^{n_{row}}\sum_{k=1}^{n_r}\sum_{p=1}^{2} x_{ijrkp} \bullet load_{rkp} \le C_{ij}, i = 1,2, j = 1,..,n_{ps}; \quad (8)$$

Equation (1) is the optimization objective of this paper, and $\boldsymbol{w_1}$ and $\boldsymbol{w_2}$ are the weights. $\boldsymbol{load_{rkp}}$ in equation (2) denotes the load of the pth route of the $\boldsymbol{k}$th rack in rth column, which can be collected in real time during the operation of the computer room. $\boldsymbol{Q_i}$ in equation (2) denotes the total load of the ith UPS. $\boldsymbol{Q_{ij}}$ in equation (3) denotes the phase $\boldsymbol{j}$ load of the $\boldsymbol{i}$th UPS. $\boldsymbol{D_i}$ in equation (4) represents the variance of load between phases of the $\boldsymbol{i}$th UPS. Equation (5) denotes the connection between each circuit of each cabinet and the bus bar above the cabinet. When the value of $\boldsymbol{x_{ijrkp}}$ is 1, the circuit of the pth route of the $\boldsymbol{k}$th rack in rth column is taken from the phase $\boldsymbol{j}$ of the $\boldsymbol{i}$th UPS, and when the value of $\boldsymbol{x_{ijrkp}}$ is 0, it represents that the circuit of the pth route of the $\boldsymbol{k}$th rack in $\boldsymbol{r}$th column is not taken from the phase $\boldsymbol{j}$ of the $\boldsymbol{i}$th UPS. Equation (6) indicates that each circuit of each cabinet can only be taken from one phase of an UPS, and that two circuits are taken from different UPS. Equation (7) indicates that the power of each rack from one UPS phase cannot exceed the capacity $\boldsymbol{C_{ijr}}$ of the bus above the cabinet. Equation (8) denotes that the power of all phases taken from an UPS in a computer room cannot exceed the capacity $\boldsymbol{C_{ij}}$ of the phase of the UPS.

From the power supply structure of microbus and the optimization fomulars, it can be seen that each rack can get power from phase A, B and C of this column separately, and only from one phase. As from which phase to get power can make UPS output more balanced, this kind of problem is a kind of integer programming problem [6], which is NP-hard problem [7]. The most effective solution to this kind of problem is branch and bound algorithm [8].

4.2 Branch and Bound Based Algorithm to Optimize The Power Distribution in Microbus

Branch and bound algorithm is a method of searching the solution of the problem by using the solution space tree of the problem. The specific method is to decompose the given problem into several relatively small sub-problems (called branches), and each sub-problem can continue to decompose downward until the decomposed sub-problems can no longer continue to decompose downward or the sub-problems cannot produce optimal solutions. In the process of branching the problem to be solved, it is called bounding to estimate the corresponding bounds or target ranges for each subproblem decomposed. The purpose of delimitation is to predict the trend of approaching the target value according to the characteristics of the problem, and to retain the branch whose trend is correct or cannot be judged yet. Deleting branches that deviate significantly from the target and tend to or are sure to be the optimal solution is called pruning. Through this method, the number of operations can be simplified and the convergence can be accelerated.

There are many constraints in power distribution system, which can be divided into three parts:

(1) The first part is that the total cabinet load of each distribution bus does not exceed the carrying capacity of the bus.

(2) The second part is load balancing among three-phase UPS.

(3) The third part is load balancing between two parallel dual bus UPS.

Therefore, according to the above constraints, a solution space tree can be established to solve the problem. Firstly, the solution space tree is established for the first and second constraints.

Each rack needs and only accesses a bus, that is, the rack is powered by a single-phase bus. The microbus composes three-phase buses, and each phase electricity is a bus, expressed by $\boldsymbol{r_x(P), x \in N, P \in (A, B, C)}$, for example, phase $\boldsymbol{A}$ of $\boldsymbol{r_1}$ column is represented by $\boldsymbol{r_1(A)}$. Each bus is equivalent to a knapsack, and the cabinet is regarded as the object to be put into the knapsack. It is represented by $\boldsymbol{r_x^{cy}, x \in N, y \in N}$, for example, $\boldsymbol{r_1^{c1}}$ represents the first rack in the $\boldsymbol{r_1}$ column. For each parent row of each column, a solution space tree is established. The solution space tree for each phase bus of column 1 is as figure 3:

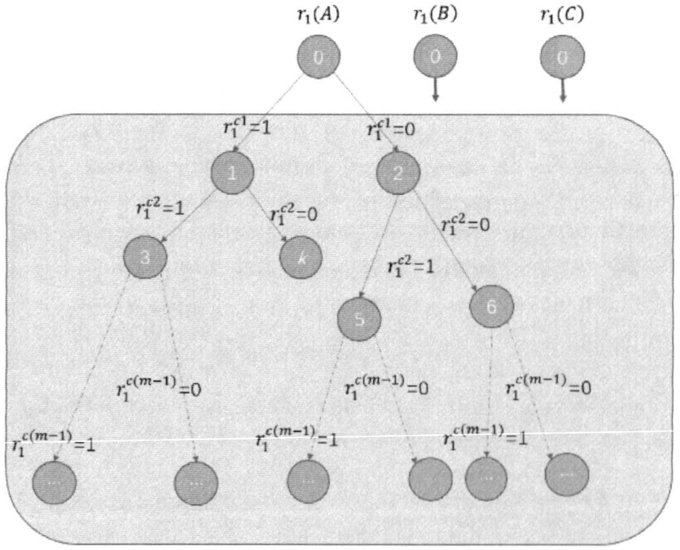

Figure 3. Solution spatial tree of power supply buses.

The green node in figure 3 represents each computing branch. When the value of the objective function of computing branch $\boldsymbol{k}$ is larger than that of its parent node, the branch is cut off and its sub-node branches are no longer calculated. $\boldsymbol{r_x^{cy}} = \mathbf{1}$ indicates that the cabinet $\boldsymbol{cy}$ is connected to the corresponding busbar of the $\boldsymbol{r_x}$column, and $\boldsymbol{r_x^{cy}} = \mathbf{0}$ indicates that the cabinet is not connected. The objective function value of each branch node is calculated sequentially until the smallest branch node, then the algorithm ends and which bus is the preferable power supply for each cabinet.

5. Experiments and discussion

In the computer room, it is usually powered by several UPSs, and each UPS supply power to two or four columns of racks according to its capacity. In this paper, according to the real power distribution system of a data center, modeling and simulation are carried out. There are 2 UPSs supplying power to 3 columns of racks. Each column is composed by 12 racks. Other parameters are set as table 1:

Table 1. Parameters of power supply UPSs.

Serial number of UPS	Total rated capacity(kVA)	Available capacity(kVA)	Phases	Rated capacity of each phase(kVA)	Number of racks	Power supply routes
ups1	350	280	A/B/C	35.2/35.2/35.2	36	2
ups2	350	280	A/B/C	35.2/35.2/35.2	36	2

Three columns of racks with traditional power supply are selected as well as their current loads and power sources are recorded as table 2:

Table 2. Load of each rack.

Power supply and load		c1	c2	c3	c4	c5	c6	c7	c8	c9	c10	c11	c12
r1_1 route	load(kVA)	2.12	1.34	4.2	2.3	2.13	3.5	4.11	1.4	0.8	4.43	1.32	1.87
	get power from	UPS1 A	UPS1 B	UPS1 C	UPS1 A	UPS1 B	UPS1 C	UPS1 A	UPS1 B	UPS1 C	UPS1 A	UPS1 B	UPS1 C
r1_2 route	load(kVA)	1.23	2.34	0.56	1.06	2.1	0.67	1.45	3.28	1.21	0.23	1.59	1.45
	get power from	UPS2 A	UPS2 B	UPS2 C	UPS2 A	UPS2 B	UPS2 C	UPS2 A	UPS2 B	UPS2 C	UPS2 A	UPS2 B	UPS2 C
r2_1 route	load(kVA)	2.13	1.34	3.2	2.2	2.23	3.5	3.51	1.1	1	2.43	4.02	1.89
	get power from	UPS1 A	UPS1 B	UPS1 C	UPS1 A	UPS1 B	UPS1 C	UPS1 A	UPS1 B	UPS1 C	UPS1 A	UPS1 B	UPS1 C
r2_2 route	load(kVA)	1.03	2.34	0.46	1.34	2.12	0.67	1.45	3.25	2.01	2.23	2.79	1.45
	get power from	UPS2 A	UPS2 B	UPS2 C	UPS2 A	UPS2 B	UPS2 C	UPS2 A	UPS2 B	UPS2 C	UPS2 A	UPS2 B	UPS2 C
r3_1 route	load(kVA)	2.12	1.24	4.2	2.3	0.13	2.5	3.61	1.42	1.8	3.43	2.22	1.61
	get power from	UPS1 A	UPS1 B	UPS1 C	UPS1 A	UPS1 B	UPS1 C	UPS1 A	UPS1 B	UPS1 C	UPS1 A	UPS1 B	UPS1 C
r3_2 route	load(kVA)	1.23	2.34	0.56	2.56	1.1	1.07	0.45	3.28	1.21	1.03	1.19	1.08
	get power from	UPS2 A	UPS2 B	UPS2 C	UPS2 A	UPS2 B	UPS2 C	UPS2 A	UPS2 B	UPS2 C	UPS2 A	UPS2 B	UPS2 C

Through table 2, the traditional UPS load can be calculated as table 3:

IOP Conf. Series: Materials Science and Engineering **752** (2020) 012008 doi:10.1088/1757-899X/752/1/012008

Table 3. Unoptimized UPS load.

Serial number of UPS	Phase	Load(kVA)	Imbalance ratio(%)
	A	34.69	22.94
UPS1	B	19.89	29.51
	C	30.07	6.57
	A	15.29	17.22
UPS2	B	27.72	50.08
	C	12.4	32.86

The load of UPSs optimized by the proposed algorithm in this paper is as table 4:

Table 4. Optimized UPS load.

Serial number of UPS	Phase	Load(kVA)	Imbalance ratio(%)
	A	23.29	0.23
UPS1	B	23.29	0.23
	C	23.45	0.46
	A	23.28	0.27
UPS2	B	23.39	0.20
	C	23.36	0.07

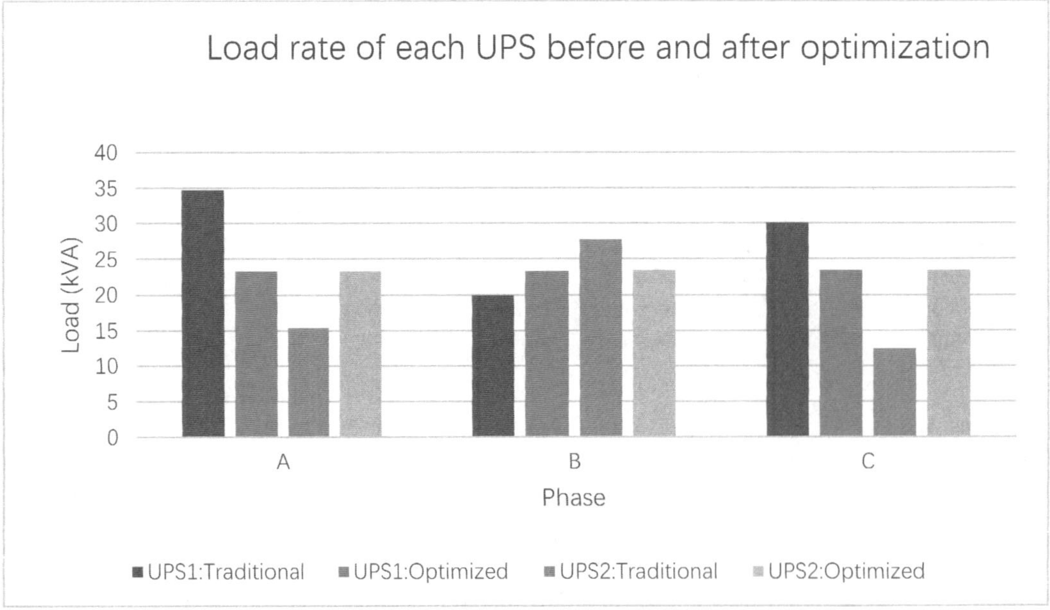

Figure 4. Comparison of UPS load before and after optimization

IOP Conf. Series: Materials Science and Engineering **752** (2020) 012008 doi:10.1088/1757-899X/752/1/012008

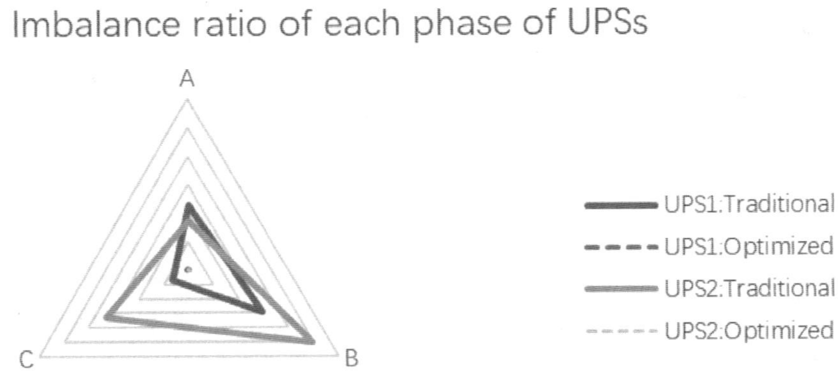

Figure 5. Imbalance ratio of each phase of UPSs before and after optimization

As can be seen from table 4, figure 4 and figure 5, the load of each UPS that is optimized is equal, which is 70.03 kVA, and the load of each phase of UPSs is balanced. It shows that the microbus power supply mode is superior to the traditional power supply mode in UPS load balancing and power distribution.

6. Conclusion

In this paper, the characteristics of traditional distribution system in data center are analyzed. It is pointed out that the microbus power supply mode can promote the three-phase balance of power supply, improve the efficiency of cabinet utilization and reduce the investment cost, which is of great help to the economic and safe operation of power distribution system. At the same time, a control strategy of rack power supply based on branch and bound algorithm is presented. By modeling and simulating the actual power consumption of a data center computer room, it can be seen that the algorithm in this paper can greatly balance the balance of three-phase power supply, thus reducing the potential safety hazards of distribution system.

Although the power distribution of bus-type power supply system is deeply studied in this paper, there are many problems involved in the computer room, so further research is needed to give full play to the advantages of bus-type power supply system.

(1) Busbar power supply system can be adjusted online according to the actual power consumption of the cabinet. The reading of the electrical parameters and the release of the control strategy need an advanced power environment monitoring system to assist in the completion.

(2) When the size of the computer room is large, the cabinet is required to adjust the bus automatically according to the downward control strategy. Therefore, the cabinet side needs to cooperate with the automatic conversion device.

References

[1] 13th Five-Year Plan for energy and resources conservation in public institutions by national government offices administration of China. http://ecpi.ggj.gov.cn/news/56166 [accessed 8 October 2019].

[2] Microbus model of power distribution by STARLINE company. https://www.starlinepower.com/busway. [accessed 8 October 2019].

[3] Elmer Sorrentino, Juan Carlos Burgos. Rules to estimate the expected values of zero-sequence impedances in 3-phase core-type transformers. 2018 *Electric Power Systems Research.* **165** 94-101

[4] Sunghyok Kim, Songchol Hyon, Chonung Kim. Distributed virtual negative-sequence impedance control for accurate imbalance power sharing in islanded microgrids. 2018 *Sustainable Energy, Grids and Networks.* **16** 28-36

[5] Fang Cao, Yajing Wang, Feng Zhu, Yujie Cao, Zhaohao Ding. UPS node based workload management for data centers considering flexible service requirements. In: *2019 IEEE/IAS 55th Industrial and Commercial Power Systems Technical Conference*. (Calgary, Canada) p 1-9.
[6] Reed Harder, Vikrant Vaze. An integer programming approach to fisheries observer deployment. 2019 *Transportation Research Part E* **127** 132-149
[7] Xiaodong Wang, Jun Tian. Dynamic programming for NP-Hard problems. 2011 *Procedia Engineering* **15** 3396-3400
[8] Jeffrey Schaller, Jorge Valente. Branch-and-bound algorithms for minimizing total earliness and tardiness in a two-machine permutation flow shop with unforced idle allowed. 2019 *Computers & Operations Research* **109** 1-11

IOP Conf. Series: Materials Science and Engineering **752** (2020) 012009 doi:10.1088/1757-899X/752/1/012009

Thrust Estimation for Aero-engine Based on Deep Convolution Neural Network

Binbin Wang, Jianguo Xu*, Xingyi Liu and Qiangang Zheng

Nanjing University of Aeronautics and Astronautics

*xujianguo@nuaa.edu.cn

Abstract. Aero-engines' main task is to provide thrust for military or civil aircraft, so direct thrust control technology becomes one of the important research directions in the aero-engine control field. Thrust estimator design is an important part of direct thrust control. To overcome the shortcomings of traditional thrust estimation methods, such as low accuracy and poor robustness, a deep convolution neural network (DCNN) was introduced to design thrust estimator for aero-engine. Taking a turbofan engine component-level model as the objective, 100000 data samples were generated through simulation for training and testing the estimator. Twelve measurable parameters, such as fuel flowrate (W_f), combustion chamber exit temperature (T_4), high pressure rotor speed (HPS_N_{mech}), low pressure rotor speed (LPS_N_{mech}), nozzle exit temperature (T_8), etc. were selected as the input features of the estimator, and engine thrust was used as the output label. The effects of training data and test data partition ratio, convolution and pooling layer number, convolution kernel size, pooling window size and channel number of convolution/pooling layers on the performance of the thrust estimator were compared and analysed. Furthermore, the performance of the DCNN based thrust estimator was compared with that of traditional thrust estimators based on BP neural network and support vector machine (SVM). The validation results show that the DCNN based thrust estimator proposed in this paper has excellent thrust estimation performance, and its average accuracy can reach as high as 0.11%.

1. Introduction

With the increasing attention paid to the safety, economy and environmental friendliness of aircraft, the requirements for aero-engine control system are getting higher and higher. The traditional aero-engine control methods are based on the closed-loop control of the measurable parameters such as rotor speed or pressure. But for turbojet or turbofan engines, the engine thrust is the parameter that people really care about when the safety is guaranteed [1-6]. Because it is very difficult to directly measure the thrust of an engine, the usual methods are to indirectly reflect the thrust using other parameters such as rotor speed, pressure and temperature. However, the relationship between these measurable parameters and the thrust is complex, and tends to change when the engine state changes, resulting in the problem that the engine thrust cannot be accurately controlled.

Scholars have done much relevant research on thrust estimator design technology. Zhou Jun et al. proposed a EW-ELM algorithm for thrust estimation, which was designed based on extreme learning machine (ELM) and wavelet theory [7]. Yao Yanlong et al. presented an adaptive genetic neural network algorithm (AGNNA), which has good global convergence property and rapid local searching capability, and can predict aero-engine thrust in the full flight envelope [8]. Maggiore et al. designed a

Content from this work may be used under the terms of the Creative Commons Attribution 3.0 licence. Any further distribution of this work must maintain attribution to the author(s) and the title of the work, journal citation and DOI.
Published under licence by IOP Publishing Ltd

thrust estimator by using the non-linear mapping ability of neural network [9]. However, the structure of shallow neural network is very simple, and the initial weight matrix is usually generated randomly, so it is prone to fall into local optimal solution. Deep convolution neural network (DCNN) has excellent performance in feature extraction and modelling and has been more and more studied and applied in academia and industry in recent years. In this paper, DCNN was studied and used to design the thrust estimator for aero-engine.

2. DCNN Model

2.1. DCNN

DCNN is constructed by imitating the biological visual perception mechanism, and can be used for supervised learning and unsupervised learning. The sharing of convolution kernel parameters and sparsity of interconnection between layers in hidden layers enable DCNN to extract features with small computation. The basic structure of DCNN is composed of an input layer, several convolution layers, several pooling layers, a full connection layer and an output layer. The convolution and pooling layers are alternately arranged and connected. The last pooling layer is reshaped as the full connection layer, and the output layer is determined according to the type of outputs [10].

There are two important concepts in DCNN convolution layer: local receptive fields and shared weights. Assuming that the input is shown in Figure 1, the red-line area in the figure is the local receptive field. The neurons in the hidden layer are only connected with 3*3 neurons in the input layer. In other words, each neuron in the hidden layer has a fixed-size visual field to observe the features of the previous layer.

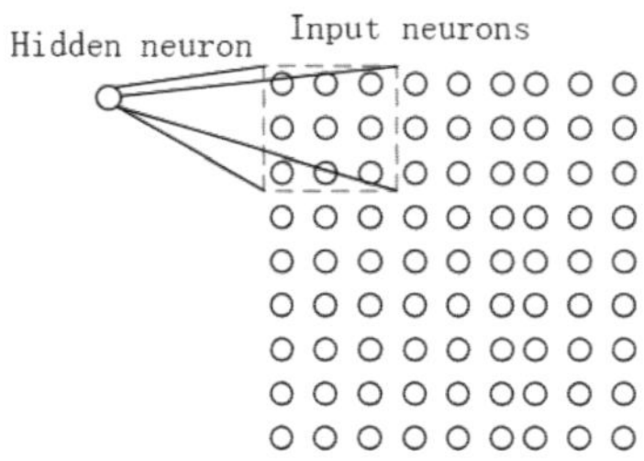

Figure 1. Local receptive field of a hidden neuron

Each local receptive field has a weight matrix, convolution kernel, which is the same size as the field. If each convolution kernel is different from each other, the computation will be very large in the training process. So a common practice is that different local receptive fields share the same convolution kernel.

The pooling layer is used to reduce the dimension of the feature maps outputted from the convolution layer. Similarly, the pooling layer also has a pooling window to scan the feature map. And there are two main types of pooling: maximum pooling and average pooling.

2.2. Model Construction

As shown in Table 1, twelve typical measurable parameters of turbofan engine were selected as the input features of the thrust estimator, including fuel flowrate (W_f), combustion chamber exit temperature (T_4), high pressure rotor speed (HPS_N_{mech}), low pressure rotor speed (LPS_N_{mech}), nozzle exit temperature (T_8), etc. In addition, engine thrust (F_g) is taken as the output label.

Table 1. Thrust estimator parameters of turbofan engine

Symbol	Parameter
W_f	fuel flowrate
$HPC\text{-}P_r$	high pressure compressor pressure ratio
$LPC\text{-}P_r$	low pressure compressor pressure ratio

P_4	combustion chamber exit pressure
T_4	combustion chamber exit temperature
$HPT\text{-}P_r$	high pressure turbine pressure ratio
$LPT\text{-}P_r$	low pressure turbine pressure ratio
$HPS\text{-}N_{mech}$	high pressure rotor speed
$LPS\text{-}N_{mech}$	low pressure rotor speed
P_{45}	high pressure turbine exit pressure
T_{45}	low pressure turbine exit temperature
T_8	nozzle exit temperature
F_g	engine thrust

DCNN usually uses square matrices as input. Wei Zhang, etc. used one-dimensional tensors as input data, and the network training efficiency was improved [11]. We used the same method to input the feature parameters. The input data first passes through the convolution layer, which scans the input data with a certain stride. The output of the convolution layer acts as the input of the next pooling layer, and the pooling layer scans the input data with the pooling window in a fixed stride. There may be several convolution layers and pooling layers. The last pooling layer is reshaped as the full connection layer, and the output layer produces engine thrust prediction. The structure of the DCNN based thrust estimator is shown in Figure 2.

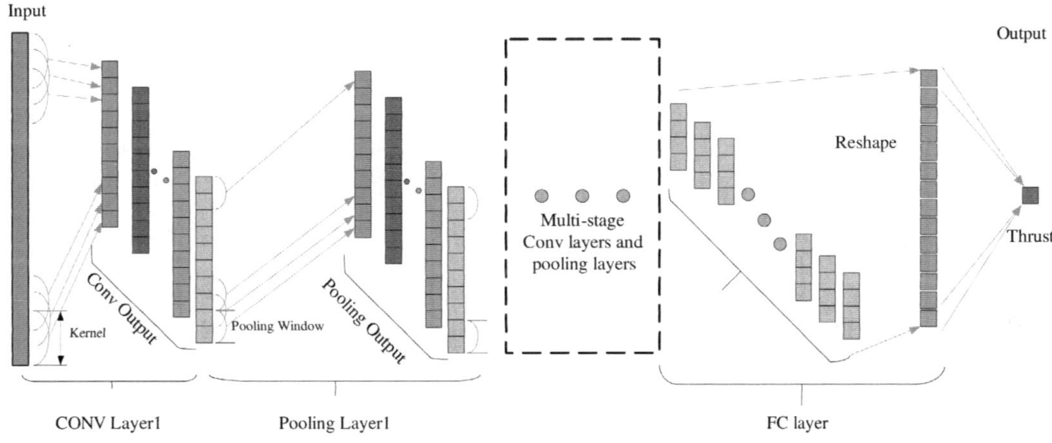

Figure 2. Structure of DCNN based thrust estimator

3. Performance analysis

The dataset used for training thrust estimator in this paper were all generated through simulation with a component-level model of a turbofan engine [12]. The dataset has 100,000 data samples which can be chosen as either training samples or test samples.

Because the order of magnitude of each input feature is quite different, some training problems such as gradient disappearance may occur during the training process and the model accuracy will be greatly affected. So it is necessary to normalize all the input features. We adopt the min-max normalization method as shown in formula (1).

$$x^* = \frac{x - x_{min}}{x_{max} - x_{min}} \tag{1}$$

In the formula (1), x_{min} and x_{max} denote the minimum and the maximum of the input feature x, respectively, and x^* denotes the normalized value of x. Thus the feature x is re-scaled within the range [0, 1].

3.1. training data and test data partition ratio

To demonstrate and illustrate the performance of the DCNN thrust estimator model, it was compared with two other AI approaches, backpropagation neural network (BP) and support vector machine (SVM) [13]. BP neural network adopts the structure of single hidden layer with 6 neurons. SVM uses Gauss kernel function, the regularization factor C is 100, the width of Gauss kernel γ is 3, and the error coefficient is 0.01. The learning rate for the three methods is set to be 0.01, and the training iteration number is 100.

The original dataset was divided into training sets and test sets according to the ratio of 8:2, 6:4 and 5:5. And the thrust estimation accuracy for the three methods, which is represented by mean square error (MSE), is shown in Figure 3.

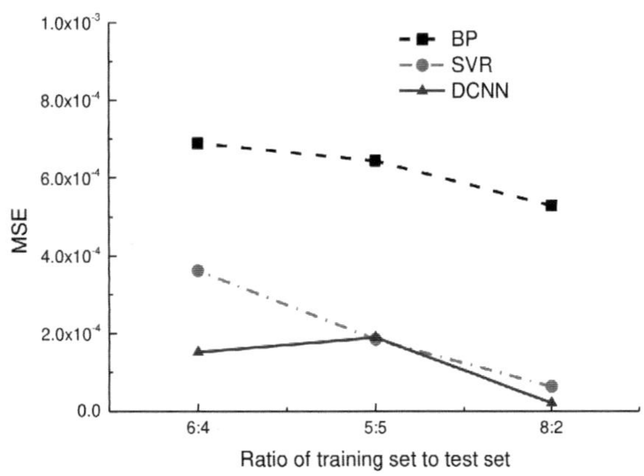

Figure 3. Performance comparison for DCNN, BP and SVR models with different training data and test data partition ratio

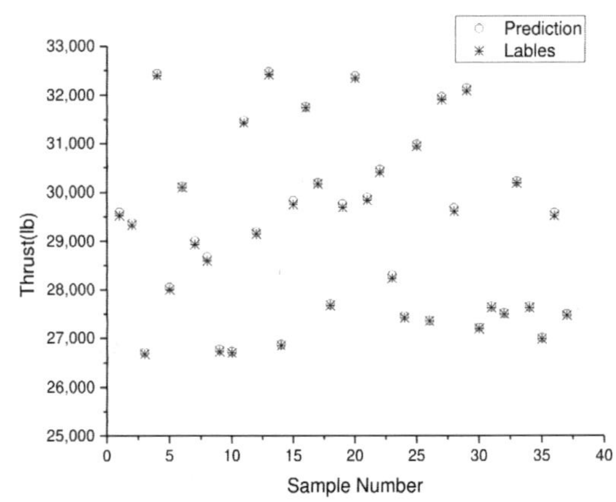

Figure 4. Comparison between DCNN prediction values and thrust labels when the training data and test data partition ratio is 8:2

As seen in Figure 3, the DCNN model is more accurate than the other two models, and when the ratio of training set to test set is 8:2, the performance of the three models all become best. So the training data and test data partition ratio is set at 8:2 in this paper. Figure 4 shows some of comparison results between DCNN prediction values and thrust labels.

3.2. Number of convolution/pooling layers

For DCNN, there are two different types of hidden layers: convolution layer and pooling layer. Different DCNNs can be constructed with different number of convolution layers and pooling layers. Although the increase of hidden layers enhances the ability to extract features, more hidden layers does not mean better performance for specific applications. In addition, because of the different effects of convolution layer and pooling layer, the performance of different DCNN structures is usually not same. Suppose Conv represents convolution layer, Pool represents pooling layer and Full represents full connection layer. The performace of four different network structures, Conv-Full, Conv-Pool-Full, Conv-Conv-Full and Conv-Pool-Conv-Pool-Full, were compared with the channel number set to 16. The comparison results are shown in Table 2.

Table 2. MSE of different DCNN Structures

CNN Structure	MSE	Max MSE	Min MSE
Conv-Full	$1.82*10^{-4}$	$1.5*10^{-2}$	$1.79*10^{-5}$
Conv-Pool-Full	$8.36*10^{-6}$	$6.39*10^{-4}$	$1.19*10^{-6}$
Conv-Conv-Full	$4.42*10^{-5}$	$3.31*10^{-3}$	$2.03*10^{-6}$
Conv-Pool-Conv-Pool-Full	$2.07*10^{-5}$	$1.76*10^{-3}$	$1.53*10^{-6}$

Table 2 shows that when the network structure is Conv-Pool-Full, the thrust estimation accuracy is the highest, and the average MSE is $8.36*10^{-6}$. And as the number of hidden layers increases, the MSE increases accordingly. Although the MSE of the last network structure also tends to decrease, the second network structure, Conv-Pool-Full, was chosen to design the thrust estimator for its simplicity.

3.3. Channel number

Channel number is usually determined based on experience. Academia has provided many types of DCNNs, such as LeNet-5, AlexNet, VGGNet and so on. Because LeNet model has excellent performance and is widely used, we choose the channel number referring to LeNet-5 model. We constructed 4 DCNN models which channel numbers are 6, 16, 32 and 64 respectively, and the training results of these models are shown in Table 3.

Table 3. MSE of DCNN with different channel number

Type	Conv6+Pool6	Conv16+Pool16	Conv32+Pool32	Conv64+Pool64
MSE	$2.8*10^{-5}$	$8.36*10^{-6}$	$4.37*10^{-5}$	$8.8*10^{-5}$

As seen in Table 3, the training effects are not very good if the channel numbers are too small or too large. When the channel number is 16, which is close to the number of input features, the estimation performance of DCNN is the best. Figure 5 shows the performance of the DCNN models change during the training process. As can be seen, although Conv64+Pool64 model has better performance in the initial training stage, Conv16+Pool16 model becomes best in the middle and later stages and has the best average performance.

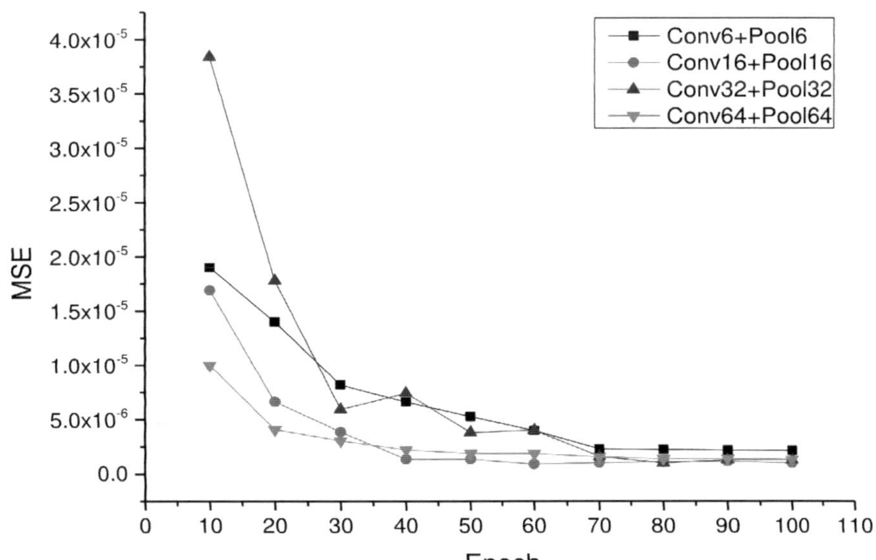

Figure 5. Training performance curves of DCNN models with different channel numbers

3.4. Convolution kernel and pooling window size

The sizes of convolution kernel and pooling window also have influence on DCNN performance and they are usually set as 2, 3 or 5 based on experience. In this paper, the DCNN models were constructed with a convolution layer and a pooling layer. So we trained and tested the DCNN models with 1*3 and 1*5 convolution kernel, and 1*2, 1*3 and 1*5 pooling window. As can be seen in Table 4, when the convolution kernel size is 1*5 and the pooling window size is 1*5, the DCNN model has the highest accuracy. The MSE of the DCNN model is $6.32*10^{-6}$ and the average accuracy is 0.11%.

Table 4. MSE of DCNN with different convolution kernel and pooling window size

Conv Kernel Size	Pooling Window Size	MSE
1*3	1*2	$8.36*10^{-6}$
1*3	1*3	$5.80*10^{-5}$
1*3	1*5	$6.69*10^{-5}$
1*5	1*2	$1.54*10^{-4}$
1*5	1*3	$2.07*10^{-4}$
1*5	1*5	$6.32*10^{-6}$

4. Conclusions

In this paper, an aero-engine thrust estimator was designed based on DCNN. The training and test datasets for the estimator were generated through simulation with a turbofan engine component level model. The effects of training data and test data partition ratio, convolution and pooling layer number, convolution kernel and pooling window size, channel number, etc. on the performance of the thrust estimator were studied. The validation test results show that the DCNN based thrust estimator has excellent performance and is better than the traditional thrust estimators based on BP neural network and SVM.

References

[1] J. X. Mu, David Rees, G. P. Liu. Advanced controller design for aircraft gas turbine engines[J]. Control Engineering Practice, 2004, 13(8).
[2] Tmothy R. Conners, Robert L. Sims. Full flight envelope direct thrust measurement on a supersonic aircraft[R]. NASA/TM-1998-206560.
[3] Rupesh Kumar, Hyunkyoung Shin. Thrust estimation of a flapping foil attached to an elastic plate using multiple regression analysis[J]. International Journal of Naval Architecture and Ocean Engineering,2019,11(2).
[4] Maggiore M., Ordonez R., Passino K, et al. Estimator designin jet engine applications[C]. Proceedings of the 38th Conference on Decision and Control . 1999.
[5] Q. Jing, C. Wu. Beidou satellite maneuver thrust force estimation for precise orbit determination[J]. GPS Solutions, 2018, 22(2).
[6] O. N. Korsun, B. K. Poplavsky, S. Ju. Prihodko. Intelligent Support for Aircraft Flight Test Data Processing in Problem of Engine Thrust Estimation[J]. Procedia Computer Science, 2017, 82-87.
[7] J. Zhou, T. H. Zhang. Aero-engine Thrust Estimation Based on Ensemble of Improved Wavelet Extreme Learning Machine[J]. Transactions of Nanjing University of Aeronautics and Astronautics, 2018, 35(02): 290-299.
[8] Y. L. Yao, J. G. Sun. Applicationg of adaptive genetic neural network algorithm in design of thrust estimator[J]. Journal of Aerospace Power, 2007, 22(10): 1748-1753.
[9] Maggiore M., Ordóñez R., Passino K. M., et al. Estimator design in jet engine applications[J]. Engineering Applications of Artificial Intelligence, 2007, 16(7): 579-593.
[10] J. Gan, W. Q. Wang, K. Lu. A new perspective: Recognizing online handwritten Chinese characters via 1-dimensional CNN[J]. Information Sciences, 2019, 375-390.
[11] W. Zhang, G. L. Peng, C. H. Li, Y. H. Chen, Z. J. Zhang. A New Deep Learning Model for Fault Diagnosis with Good Anti-Noise and Domain Adaptation Ability on Raw Vibration Signals [J]. Sensors (Basel, Switzerland), 2017, 17(2).
[12] J. W. Chapman, R. D. May, T. M. Lavelle, et al. Toolbox for the Modeling and Analysis of Thermodynamic Systems (T-MATS) User's Guide[J]. 2014.
[13] Y. P. Zhao. Support vector regression and their applications to parameter estimation for intelligent aero-engines [D]. Nanjing: Nanjing University of Aeronautics and Astronautics, 2009. (in Chinese)

IOP Conf. Series: Materials Science and Engineering **752** (2020) 012010 doi:10.1088/1757-899X/752/1/012010

Assessment of the Voltage Influence of Charging Electrical Vehicles on the Distribution Grid

Rui Ye*, Xueliang Huang and Ran Duan

School of Electrical Engineering, Southeast University, Nanjing 210096, China

*yerui8500@126.com

Abstract. At present, most relevant researches study the impact of charging electrical vehicles (EVs) on distribution network through simulation method. This method requires a large amount of input information and simulation calculation, and requires a large amount of simulation time. The required input information is difficult to be fully satisfied in the actual situation, which is easy to influence the evaluation effect. In order to overcome the drawbacks of the simulation method, an assessment model for evaluating the voltage influence caused by EV charging on distribution grid has been proposed in this paper, which can work with only the traditional distribution network input information for load flow calculation and historical charging load information at charge charging stations(FCS). The proposed model can perform early warning type assessment of voltage change, which can assist the distribution grid operators to find trouble nodes with high probability of voltage violation.

1. Introduction

In recent years, with the rapid development of EVs, the impact of charging EVs on the power grid is increasingly prominent. EVs usually access the low voltage distribution grid, making it easier to encounter transformer overload, voltage problems, cable overload, etc. [1]. Therefore, it is practically significant to research the impact of charging EVs on distribution grid.

Recently, there have been numerous related work. [2] proposes that voltage problems may occur before cable overload. In [3], an integrated algorithm based on Monte Carlo Simulation (MCS) and Time-Series analysis for evaluating EV's impact on power grid has been proposed, and the research finds that more quickly the charging load changes, faster the transformer tapping operating frequency increases, which may greatly reduce the transformer life and significantly increase the risk of voltage violation. With the augment of EV's penetration, distribution network investment and energy losses would increase obviously, and when EV's penetration reaches 62%, energy losses can add up to 40% of actual values in off-peak hours [4]. [5] assesses the impact of EVs charging on distribution grid, which centres on several distribution grid parameters, such as transformer load, feeder load, voltage deviation and energy losses, and obtains the conclusion that 60% EV's penetration would cause 64.9% load increase with uncoordinated charging through simulation. In [6], authors analyse total power losses and maximum voltage deviation by using the simulation method, and the coordinated charging could effectively decrease the impact of EVs charging on total power losses and voltage deviation. Approaches used to produce EV charging load in above research generally predetermine the charging allocation, which makes the research results more strongly related to the predetermination. To reduce the relationship between research results and predetermination, a probabilistic charging demand model based on random trip chain and Markov decision process is employed to research the temporal-spatial

Content from this work may be used under the terms of the Creative Commons Attribution 3.0 licence. Any further distribution of this work must maintain attribution to the author(s) and the title of the work, journal citation and DOI.
Published under licence by IOP Publishing Ltd

characteristics of EV charging load in [7]. And in [8], the model is used to assess the effect of EV's charging on nodal voltage deviation in several simulation scenarios. For most relative work, the complex simulation method is mainstream, which mainly contains two steps as follows:

- Simulate EV's charging behaviour to produce the charging load by using predetermination method or random method.
- Put the charging load information into power grid assessment model, then using traditional power flow method to calculate the concerned power grid parameters, such as voltage amplitude.

The two above steps all need spend a lot of time on calculation, moreover, either predetermination method or random method needs numerous input information gotten difficultly in practical. If we missed any input information in application, the simulation results might have a large deviation in comparison to the actual situation, thus might mislead decision maker's judgment.

Some direct sensitivity analysis methods are proposed in [9-11], which try to avoid excessive simulation calculation. [12] proposes a novel concept of probabilistic voltage sensitivity analysis (PVSA) to depict voltage change with a rigid precondition that the random load variation must follow normal distribution with mean 0, which is not the common case, especially for EV load. In this paper, we focus on evaluating the nodal voltage change caused by EV charging load and propose a PVSA-based assessment model to evaluate the nodal voltage change, which could accommodate the random character of EV charging load and be advantageous to decrease the calculation time. Furthermore, only common historical EV charging load data at fast charging station (FCS) and traditional input information used to calculate the power flow are needed.

The paper is arranged as follows. Section 2 builds a linear approximately distribution grid model and proposes the assessment model. Section 3 demonstrates the assessment model proposed in this paper through IEEE 33 node test system. Finally, section 4 draws conclusions from the research.

2. Modelling and methodology

2.1. Linear approximate distribution grid model

Generally, the whole regional low voltage distribution grid contains several source nodes. Because of the open loop operation characteristic, it could be divided into several independent radial distribution grids. Thus, it is reasonable to research a typical radial distribution grid instead. The node where the FCS is will be regarded as FCS node, and the node to be assessed is regarded as observation node. Assume that there is only one FCS, voltage at observation node can be written as:

$$V_o = V_s - \sum_{e \in \mathbf{E}_o} I_e \cdot Z_e \tag{1}$$

where I_e is the current through the edge e, V_s is the voltage at slack node, Z_e is the impedance of edge e, and $\mathbf{E}_o$ is the set of edges between slack node and observation node. Since I_e is equal to $\sum_{\in \mathbf{N}_e}^{n}(-S_n/V_n^*)$, where V_n^* is the complex conjugate voltage at node n, $\mathbf{N}_e$ is the set of nodes n where edge e is between slack node and node n, S_n is the injected complex power at node n, then Equation (1) can be rewritten as:

$$V_o = V_s + \sum_{e \in \mathbf{E}_o} \left(\sum_{n \in \mathbf{N}_e} \frac{S_n}{V_n^*} \right) \cdot Z_e \tag{2}$$

Assume that the injected complex power of all nodes n changes from S_n to $S_n^{'}$, the corresponding nodal voltage changes from V_n^* to $V_n^{'}$, and the voltage at observation node changes from V_o to $V_o^{'}$, then the new $V_o^{'}$ can be given by

$$V_o^{'} = V_s + \sum_{e \in \mathbf{E}_o} \left(\sum_{n \in \mathbf{N}_e} \frac{S_n^{'}}{\left(V_n^{'}\right)^*} \right) \cdot Z_e \tag{3}$$

Rewrite $S_n^{'}$ and $V_n^{'}$ in increment form as follows:

$$\begin{cases} S_n^{'} = S_n + \Delta S_n \\ V_n^{'} = V_n + \Delta V_n \end{cases} \tag{4}$$

Then, based on Equation (4) and Equation (3), voltage change ΔV_o at observation node can be written as:

$$\Delta V_o = V_o^{'} - V_o = \sum_{e\in \mathbf{E}_o}\left(\sum_{n\in \mathbf{N}_e}\frac{S_n^{'}}{V_n^{'*}}\right)\cdot Z_e - \sum_{e\in \mathbf{E}_o}\left(\sum_{n\in \mathbf{N}_e}\frac{S_n}{V_n^{*}}\right)\cdot Z_e = \sum_{e\in \mathbf{E}_o}\left(\sum_{n\in \mathbf{N}_e}\frac{\Delta S_n\cdot V_n^{*} - S_n\cdot \Delta V_n^{*}}{V_n^{*}\cdot\left(V_n^{*}+\Delta V_n^{*}\right)}\right)\cdot Z_e \tag{5}$$

Since $\Delta V_n^{*} << V_n^{*}\cdot(V_n^{*}+\Delta V_n^{*})$ in common practical application, thus Equation (5) can be approximated to the following form:

$$\Delta V_o \approx \sum_{e\in \mathbf{E}_o}\left(\sum_{n\in \mathbf{N}_e}\frac{\Delta S_n\cdot V_n^{*}}{V_n^{*}\cdot\left(V_n^{*}+\Delta V_n^{*}\right)}\right)\cdot Z_e \tag{6}$$

If the injected complex power of the FCS node changes from S_d to $S_d+\Delta S_d$, then voltage change at the observation node can be written as:

$$\Delta V_{od} = \frac{\Delta S_d}{V_d^{*} + \Delta V_{dd}^{*}}\cdot Z_{od} \tag{7}$$

where V_d^{*} is complex conjugate voltage at the FCS node, ΔV_{dd}^{*} is complex conjugate voltage change caused by the injected complex power change of the FCS node, Z_{od} is shared impedance between the FCS node and observation node from slack node, which is illustrated in figure 1.

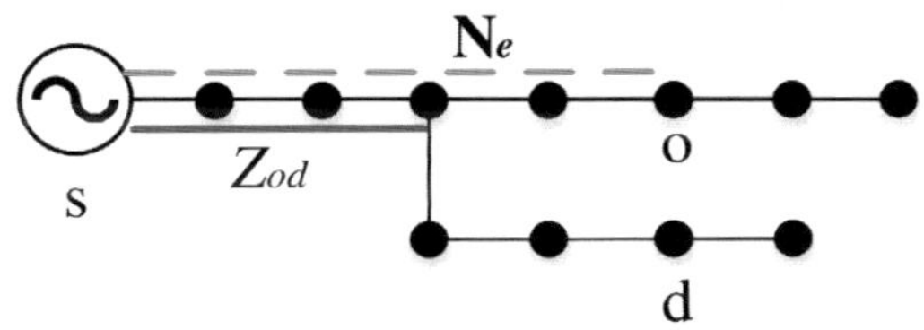

Figure 1. Shared impedance diagram.

Let $\Delta S_d=\Delta P_d+i\Delta Q_d$, $\Delta V_{od}=\Delta V_{od}^{r}+i\Delta V_{od}^{i}$, $V_d=V_d^{r}+iV_d^{i}$, and $Z_{od}=R_{od}+iX_{od}$, then we can get the equation as follows [12]:

$$\begin{cases} \Delta V_{od}^{r} \le \left[\dfrac{\left(\Delta P_d\cdot R_{od} - \Delta Q_d\cdot X_{od}\right)\cdot V_d^{r}}{\left(V_d^{r}\right)^2+\left(V_d^{i}\right)^2} - \dfrac{\left(\Delta P_d\cdot X_{od} + \Delta Q_d\cdot R_{od}\right)\cdot V_d^{i}}{\left(V_d^{r}\right)^2+\left(V_d^{i}\right)^2}\right] \\ \Delta V_{od}^{i} \le \left[\dfrac{\left(\Delta P_d\cdot X_{od} + \Delta Q_d\cdot R_{od}\right)\cdot V_d^{r}}{\left(V_d^{r}\right)^2+\left(V_d^{i}\right)^2} + \dfrac{\left(\Delta P_d\cdot R_{od} + \Delta Q_d\cdot X_{od}\right)\cdot V_d^{i}}{\left(V_d^{r}\right)^2+\left(V_d^{i}\right)^2}\right] \end{cases} \tag{8}$$

Let $V_d=|V_d|\angle\theta_d$ and substitute it into Equation(8), then Equation(8) can be turned into polar form as follows:

$$\begin{cases} \Delta V_{od}^{r} \le \left[\dfrac{R_{od}\cdot\cos\theta_d - X_{od}\cdot\sin\theta_d}{|V_d|}\cdot\Delta P_d - \dfrac{R_{od}\cdot\sin\theta_d + X_{od}\cdot\cos\theta_d}{|V_d|}\cdot\Delta Q_d\right] \\ \Delta V_{od}^{i} \le \left[\dfrac{R_{od}\cdot\sin\theta_d + X_{od}\cdot\cos\theta_d}{|V_d|}\cdot\Delta P_d + \dfrac{R_{od}\cdot\cos\theta_d - X_{od}\cdot\sin\theta_d}{|V_d|}\cdot\Delta Q_d\right] \end{cases} \tag{9}$$

Moreover, for the situation having more than one FCS, the total voltage change at the observation node satisfies Superposition Law [12]:

$$\begin{cases} \Delta V_o^r \le \sum_{d\in\mathbf{D}} \left[\frac{R_{od}\cdot\cos\theta_d - X_{od}\cdot\sin\theta_d}{|V_d|}\cdot\Delta P_d - \frac{R_{od}\cdot\sin\theta_d + X_{od}\cdot\cos\theta_d}{|V_d|}\cdot\Delta Q_d \right] \\ \Delta V_o^i \le \sum_{d\in\mathbf{D}} \left[\frac{R_{od}\cdot\sin\theta_d + X_{od}\cdot\cos\theta_d}{|V_d|}\cdot\Delta P_d + \frac{R_{od}\cdot\cos\theta_d - X_{od}\cdot\sin\theta_d}{|V_d|}\cdot\Delta Q_d \right] \end{cases} \tag{10}$$

where **D** is the set of FCS nodes. Equation (10) specifies the upper bounds of the real part and the imaginary part of voltage change at observation node. Since $|\Delta V_o|^2 = |\Delta V_o^r|^2 + |\Delta V_o^i|^2$, then the upper bound of $|\Delta V_o|^2$ can be written as the sum of the square of the upper bounds. The distribution network operators tend to care about the upper bound of $|\Delta V_o|^2$, which can provide early warning information. Thus, it is reasonable to rewrite the Equation (10) as:

$$\begin{cases} \Delta V_o^r = \sum_{d\in\mathbf{D}} \left[\frac{R_{od}\cdot\cos\theta_d - X_{od}\cdot\sin\theta_d}{|V_d|}\cdot\Delta P_d - \frac{R_{od}\cdot\sin\theta_d + X_{od}\cdot\cos\theta_d}{|V_d|}\cdot\Delta Q_d \right] \\ \Delta V_o^i = \sum_{d\in\mathbf{D}} \left[\frac{R_{od}\cdot\sin\theta_d + X_{od}\cdot\cos\theta_d}{|V_d|}\cdot\Delta P_d + \frac{R_{od}\cdot\cos\theta_d - X_{od}\cdot\sin\theta_d}{|V_d|}\cdot\Delta Q_d \right] \end{cases} \tag{11}$$

where ΔV_o^r and ΔV_o^i refer to the upper bounds of the real part and the imaginary part respectively. In this paper, Equation (11) is called linear approximate distribution grid model.

2.2. The assessment model of the nodal voltage change caused by EV's charging

Actually, the charging load at FCS presents strong randomness in time profile, thus, we can use random variable to depict the charging load. In this paper, we assume that the charging load at FCS follows normal distribution and the corresponding random variable can be given by

$$P_{j\kappa}^{\mathrm{ch}} \sim N\left(\mu_{j\kappa}, \sigma_{j\kappa}^2\right) \tag{12}$$

where $P_{j\kappa}^{\mathrm{ch}}$ stands for the charging load at the FCS connected to jth node in κth time profile, $\mu_{j\kappa}$ stands for the expectation of $P_{j\kappa}^{\mathrm{ch}}$, $\sigma_{j\kappa}$ stands for the standard deviation of $P_{j\kappa}^{\mathrm{ch}}$. $\mu_{j\kappa}$ and $\sigma_{j\kappa}$ can be gotten through maximum likelihood estimation method. Since the linear conversion of normal distribution is still normal distribution, Equational (12) can be rewritten as:

$$\begin{cases} P_{j\kappa}^{\mathrm{ch}} = \mu_{j\kappa} + \Delta P_{j\kappa}^{\mathrm{ch}} \\ \Delta P_{j\kappa}^{\mathrm{ch}} \sim N\left(0, \sigma_{j\kappa}^2\right) \end{cases} \tag{13}$$

where $\Delta P_{j\kappa}^{\mathrm{ch}}$ is the charging load change in κth time profile. Let $\mathbf{L}^{\mathrm{r}}$ and $\mathbf{L}^{\mathrm{i}}$ be the coefficient vectors. Ignoring the reactive power change, which is based on the reality that the power factor of the FCS is usually close to 1, and according to Equation (11), $\mathbf{L}^{\mathrm{r}}$ and $\mathbf{L}^{\mathrm{i}}$ can be written as:

$$\mathbf{L}^{\mathrm{r}} = \left[\frac{R_{o1}\cdot\cos\theta_1 - X_{o1}\cdot\sin\theta_1}{|V_1|} \cdots \frac{R_{on}\cdot\cos\theta_n - X_{on}\cdot\sin\theta_n}{|V_n|} \right]^{\mathrm{T}} \tag{14}$$

$$\mathbf{L}^{\mathrm{i}} = \left[\frac{R_{o1}\cdot\sin\theta_1 + X_{o1}\cdot\cos\theta_1}{|V_1|} \cdots \frac{R_{on}\cdot\sin\theta_n + X_{on}\cdot\cos\theta_n}{|V_n|} \right]^{\mathrm{T}} \tag{15}$$

where θ and V can be calculated using traditional Newton power flow method, when load at nth node is equal to $\mu_{n\kappa}$. For node without FCS, μ is equal to the mean load at the node. Let us define $\mathbf{\Delta P}=[\Delta P_{1\kappa},\ldots,\Delta P_{n\kappa}]$ as normal random vector and assume its elements are independent. According to Equation (10), (14) and (15), we have

$$\begin{cases}\Delta V_{o\kappa}^{r}=\left(\mathbf{L}^{r}\right)^{\mathrm{T}}\cdot\mathbf{\Delta P}^{\mathrm{T}}\sim N\left(0,\left(\mathbf{L}^{r}\right)^{\mathrm{T}}\mathbf{C}\mathbf{L}^{r}\right)\\ \Delta V_{o\kappa}^{i}=\left(\mathbf{L}^{i}\right)^{\mathrm{T}}\cdot\mathbf{\Delta P}^{\mathrm{T}}\sim N\left(0,\left(\mathbf{L}^{i}\right)^{\mathrm{T}}\mathbf{C}\mathbf{L}^{i}\right)\end{cases} \tag{16}$$

where,

$$\mathbf{C}=\begin{bmatrix}\sigma_{1\kappa}^{2} & 0 & 0\\ 0 & \ddots & 0\\ 0 & 0 & \sigma_{n\kappa}^{2}\end{bmatrix} \tag{17}$$

where $\sigma_{n\kappa}$ is zero for node without FCS. Let us define $\mathbf{\Delta V}=(\Delta V_{o\kappa}^{r},\Delta V_{o\kappa}^{i})^{\mathrm{T}}$. Due to both $\Delta V_{o\kappa}^{r}$ and $\Delta V_{o\kappa}^{i}$ are linear combination of $\Delta P_{1\kappa},\ldots,\Delta P_{n\kappa}$, then $\mathbf{\Delta V}\sim N(0,\mathbf{C}_1)$, where,

$$\mathbf{C}_1=\begin{bmatrix}\sum_{j=1}^{n}\left(L_j^{r}\right)^2\sigma_{j\kappa}^{2} & \sum_{j=1}^{n}L_j^{r}L_j^{i}\sigma_{j\kappa}^{2}\\ \sum_{j=1}^{n}L_j^{r}L_j^{i}\sigma_{j\kappa}^{2} & \sum_{j=1}^{n}\left(L_j^{i}\right)^2\sigma_{j\kappa}^{2}\end{bmatrix} \tag{18}$$

where L_j^{r} is jth element of $\mathbf{L}_r$ and L_j^{i} is jth element of $\mathbf{L}^{i}$. $\mathbf{C}_1$ can be diagonalized by eigenvalue decomposition as:

$$\mathbf{\Lambda}=\mathbf{W}^{\mathrm{T}}\mathbf{C}_1\mathbf{W}=\begin{bmatrix}\lambda_1 & \\ & \lambda_2\end{bmatrix} \tag{19}$$

where $\mathbf{W}$ is the Eigen matrix of $\mathbf{C}_1$, λ_1 and λ_2 are eigenvalues of $\mathbf{C}_1$. Let us define $\mathbf{U}=(U_1,U_2)\sim N(0,\mathbf{\Lambda})$, $U_1\sim N(0,\lambda_1)$, and $U_2\sim N(0,\lambda_2)$. Since the value of $\mathrm{cov}(U_1,U_2)$ is 0, U_1 and U_2 are independent. Because $\mathbf{U}$ is orthogonal transformation of $\mathbf{\Delta V}$, then,

$$\mathbf{U}^{\mathrm{T}}\mathbf{U}=\mathbf{\Delta V}^{\mathrm{T}}\mathbf{W}\cdot\mathbf{W}^{\mathrm{T}}\mathbf{\Delta V}=\mathbf{\Delta V}^{\mathrm{T}}\cdot\mathbf{\Delta V}=\left(\Delta V_{o\kappa}^{r}\right)^2+\left(\Delta V_{o\kappa}^{i}\right)^2=\left|\Delta V_{o\kappa}\right|^2 \tag{20}$$

Since $\mathbf{U}^{\mathrm{T}}\mathbf{U}=U_1^2+U_2^2$, then,

$$\left|\Delta V_{o\kappa}\right|^2=\left(U_1\right)^2+\left(U_2\right)^2 \tag{21}$$

Therefore, $|\Delta V_{o\kappa}|^2$ is the sum of two weighted independent chi-squared random variables, and the sum of weighted independent chi-squared random variables can be approximated by Gamma distribution [13]. The parameters of $\Gamma(\alpha,\beta)$ can be calculated as follows:

$$\begin{cases}\alpha=\dfrac{\left(E\left(\left|\Delta V_{o\kappa}\right|^2\right)\right)^2}{D\left(\left|\Delta V_{o\kappa}\right|^2\right)}\\ \beta=\dfrac{E\left(\left|\Delta V_{o\kappa}\right|^2\right)}{D\left(\left|\Delta V_{o\kappa}\right|^2\right)}\end{cases} \tag{22}$$

where $E(|\Delta V_{o\kappa}|^2)$ and $D(|\Delta V_{o\kappa}|^2)$ are expectation and variance of $|\Delta V_{o\kappa}|^2$, respectively., and can be given by

$$\begin{cases}E\left(\left|\Delta V_{o\kappa}\right|^2\right)=\lambda_1+\lambda_2\\ D\left(\left|\Delta V_{o\kappa}\right|^2\right)=2\left(\lambda_1^2+\lambda_2^2\right)\end{cases} \tag{23}$$

Then the probability of $|\Delta V_{o\kappa}|>\delta$ can be given by

$$F_{o\kappa}(\delta,\beta,\alpha)=\text{Prob}\left(|\Delta V_{o\kappa}|^2>\delta^2\right)=1-\int_0^{\delta^2}\frac{\beta^\alpha}{\Gamma(\alpha)}x^{\alpha-1}e^{-\beta x}dx \tag{24}$$

where δ is a set value in the range of 0 to 0.05. $F_{o\kappa}(\delta,\beta,\alpha)$ is the function used to estimate the voltage change at the observation node in κth time profile, which can provide an effective reference for grid operators. The larger value of $F_{o\kappa}(\delta,\beta,\alpha)$ is, the greater is the impact of EV's charging on voltage change at the observation node.

3. SIMULATION

In this paper, IEEE 33 node distribution grid test system is adopted to verify the proposed assessment model. The system reference voltage is 12.66kV and the reference capacity is 100MW. In order to simplify the complexity of simulation, only two FCSs are connected to nodes 9 and 24, respectively, as shown in figure 2, and nodes without FCS adopt constant power model. Figure 3 and figure 4 respectively show the frequency distribution diagram of FCS A and FCS B according to the 6 months' actual historical charging load data at two real FCSs. Using the maximum likelihood estimation method we can obtain the expectation and variance of the charging load variables, where charging load at FCS A obeys $N(280,108^2)$, and charging load at FCS B obeys $N(316,97^2)$. The unit of the parameters is kW.

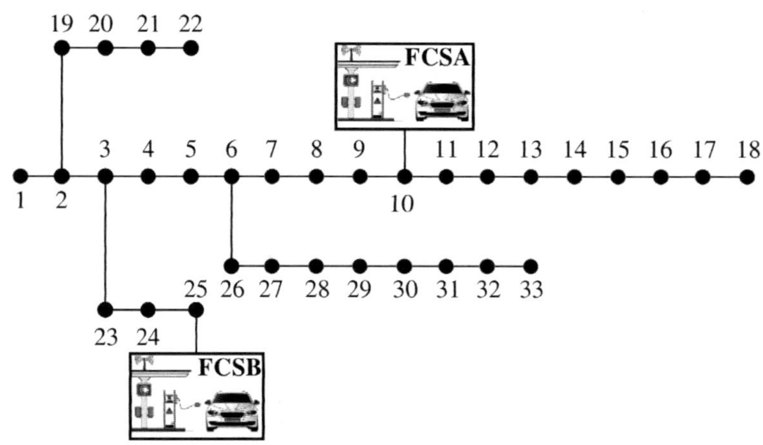

Figure 2. IEEE 33 test system.

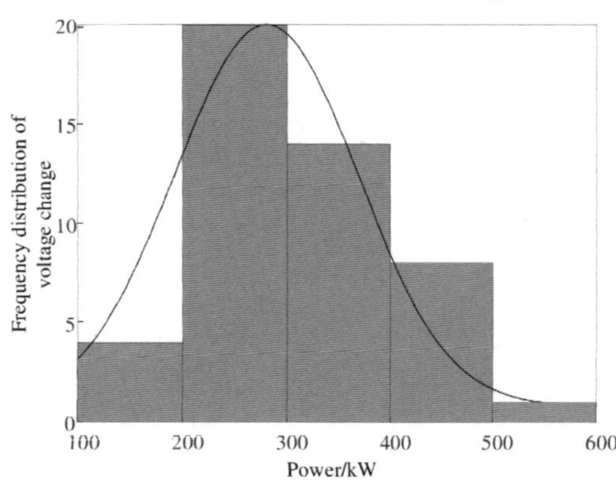

Figure 3. Frequency distribution of voltage change at FCS A.

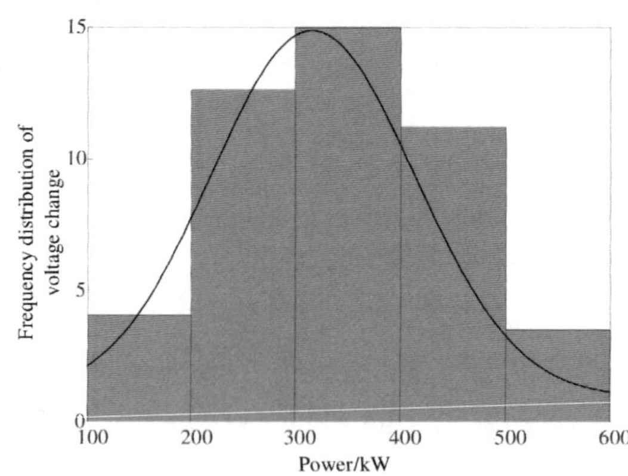

Figure 4. Frequency distribution of voltage change at FCS B.

3.1. Verification of the proposed assessment model

Figure 5 shows the probability distribution of voltage change at node 11, where the dashed line stands for the result obtained through the proposed model, as the solid line stands for the statistical result obtained by using traditional Newton power flower method. The procedure of the statistical method is as follows:

- Calculate the reference voltage amplitude at observation node through Newton power flower method, when load at FCS nodes is equal to its expectation, and load at nodes without FCS is equal to its mean value.
- Randomly generate a value obeying $N(280,108^2)$ as charging load at the FCS A, and randomly generate a value obeying $N(316,97^2)$ as charging load at the FCS B.
- Calculate the absolute value of the difference between the voltage amplitude obtained by using Newton power flower method at observation node and reference voltage amplitude.
- Iterate steps 2 and 3 until 2000 times. Then we can get probability distribution of the voltage change at observation node by using statistical analysis.

Figure 6 shows the probability distribution of voltage change at node 6. From figure 5 and figure 6 it can be seen that the result obtained through proposed assessment model is close to the result obtained through the statistical method demonstrating the effectiveness of the proposed assessment model.

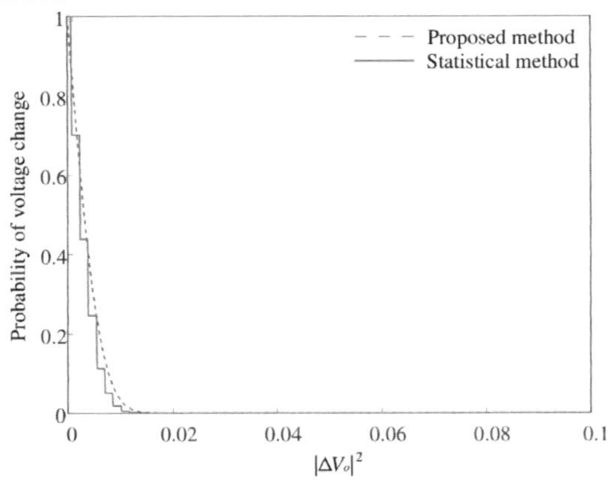

Figure 5. Probability distribution of voltage change at node 11.

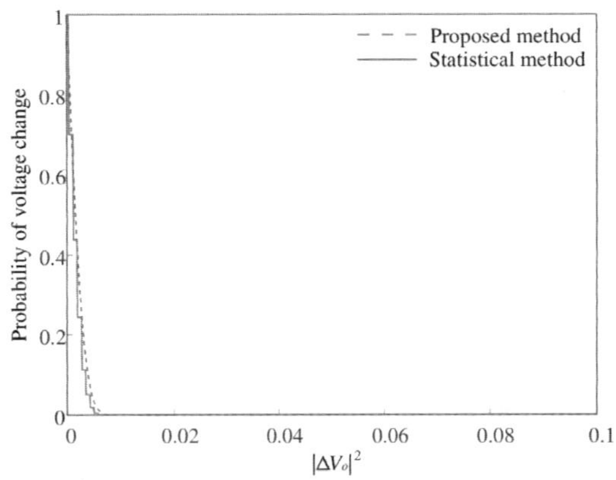

Figure 6. Probability distribution of voltage change at node 6.

3.2. Verification of the early warning characteristic

We calculate the voltage change at each node in IEEE 33 test system by using the proposed model and the statistical method respectively, and figure 7 shows the comparison between the two approaches. When δ is equal to 0.002, the overall trends of the results obtained by two approaches are similar. However, there are errors at several nodes, such as nodes 7, 9, 15, 24 and 25, which are mainly caused by the approximation of the distribution grid model, and the maximum absolute error which is close to 17% occurs at node 25. In practical application, the distribution grid operator is more concerned with δ value at 0.01~0.05. With the value of δ changing from 0.002 to 0.01, from figure 8-10, it can be seen that the voltage evaluation values obtained by the statistical method are lower than those calculated by the proposed model. If the values obtained by the statistical method are taken as the actual values, it shows that the evaluation results obtained by the proposed model are larger than they really are, which can imply the early warning information for the distribution grid operators. The distribution grid operators can pay close attention to the nodes having the assessment values out of the threshold value, and find the worst node among them to take measures to compensate the nodal voltage, for example, in figure 8, if the threshold value was 0.25, the node which might be compensated was node 15.

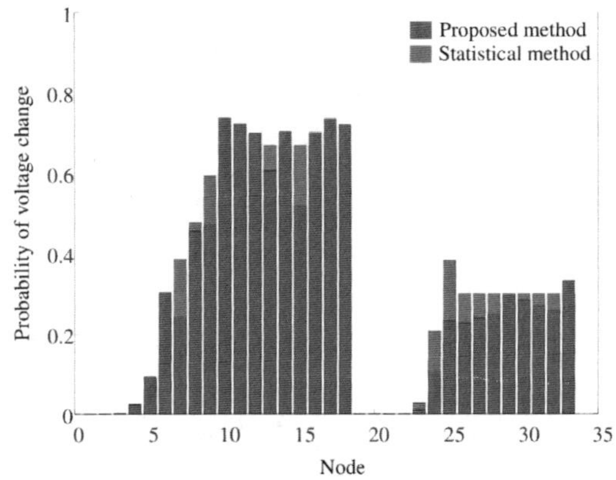

Figure 7. Probality of voltage change at each node (δ=0.002).

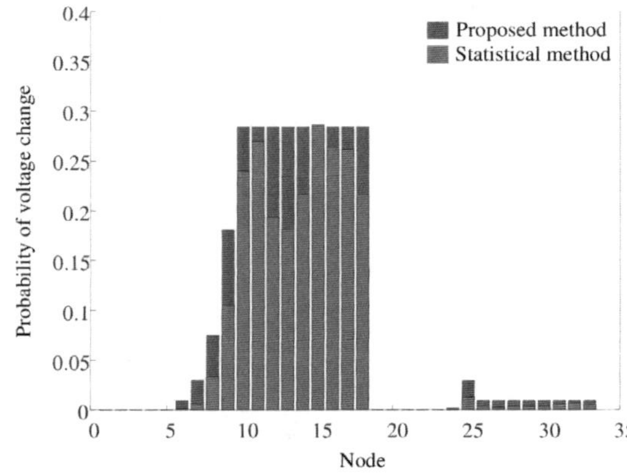

Figure 8. Probality of voltage change at each node (δ=0.005).

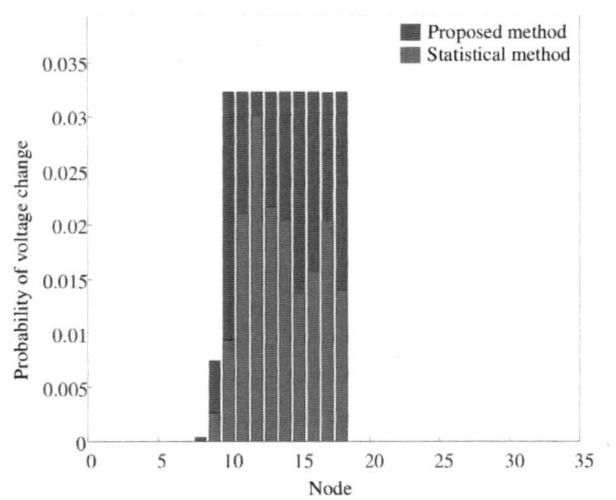

Figure 9. Probality of voltage change at each node (δ=0.07).

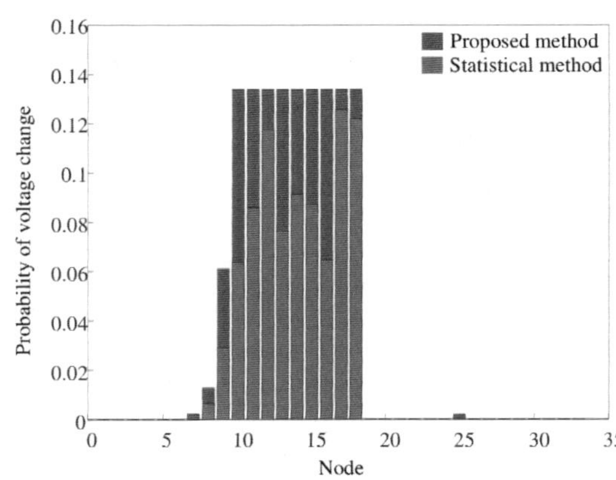

Figure 10. Probality of voltage change at each node (δ=0.01).

4. Conclusion

This paper proposes an assessment model to evaluate the impact of EV's charging on nodal voltage change. Through the establishment of the linear approximate distribution grid model, the original nonlinear distribution grid model approximate to linear model. According to the characteristic that the sum of the weighted independent chi-square random variables approximate obey Gamma distribution, we get the assessment function $F_{o\kappa}(\delta,\beta,\alpha)$. We demonstrate the effectiveness of the proposed model by using IEEE 33 node test system in this paper. The simulation results show that the probability distributions respectively calculated through proposed model and the statistical method are basically the same change trend, and the evaluation results obtained through the proposed method are usually larger than they really are, which can help the distribution grid operators find the trouble nodes in advance.

References

[1] Yunfei M, Jianzhong W, Nick J, Hongjie J, and Chengshan W 2014 A spatial-temporal model for grid impact analysis of plug-in electric vehicles *Applied Energy* **114** 456-65.

[2] Jos van B, Bernhard W and Anders A 2015 Grid impact of charging electric vehicles; study cases in Denmark, Germany and The Netherlands Powertech IEEE Eindhoven.

[3] Daijiafan M, Ziran G, Jiankang W 2019. An integrated algorithm for evaluating plug-in electric vehicle's impact on the state of power grid assets *International Journal of Electrical Power & Energy Systems* **105** 793-802.

[4] Luis P F, Tomás S R, Rafael Cossent C M D and Pablo F 2011 Assessment of the impact of plug-in electric vehicles on distribution networks *IEEE Transactions on Power Systems* **26**, 206-13.
[5] Hassanien R and Abdelfatah A 2012 Assessment of plug-in electric vehicles charging on distribution networks *IEEE Power & Energy Society General Meeting*.
[6] Kristien C N, Edwin H and Johan D 2010 The impact of charging plug-in hybrid electric vehicles on a residential distribution grid *IEEE Transactions on Power Systems* **25** 371-80.
[7] Difei T and Peng W 2015 Probabilistic modeling of nodal charging demand based on spatial-temporal dynamics of moving electric vehicles *IEEE Transactions on Smart Grid* **7** 627-636.
[8] Difei T and Peng W 2016 nodal impact assessment and alleviation of moving electric vehicle loads: from traffic flow to power flow *IEEE Transactions on Power Systems* **31** 4231-42.
[9] Zad B B, Lobry J and Valle F 2016 A centralized approach for voltage control of MV distribution systems using DGs power control and a direct sensitivity analysis method *Proc. IEEE Int. Energy Conf.*
[10] Zad B B, Hasanvand H, Lobry J and Valle F 2015 Optimal reactive power control of DGs for voltage regulation of MV distribution systems using sensitivity analysis method and PSO algorithm *Int. J.Elect. Power Energy Syst.* **68** 52–60.
[11] Zad B B, Lobry J, Vallee F and Hasanvand H 2014 Optimal reactive power control of DGs for voltage regulation of MV distribution systems considering thermal limit of the system branches *Proc. Int. Conf.Power Syst. Technol.* 2951–2958.
[12] Kumarsinh J, Balasubramaniam N and Anil PW 2018 Probabilistic voltage sensitivity analysis(PVSA)—a novel approach to quantify impact of active consumers *IEEE Transactions on Power Systems* **33** 2518-27.
[13] Li-Ling C and Y-S S 2012 Approximated distributions of the weighted sum of correlated chi-squared random variables *J. Stat. Planning Inference* **142** 457-72.

Route Planning and Charging Navigation Strategy for Electric Vehicles Based on Real-time Traffic Information and Grid Information

Xing Qiang*, Chen Zhong, Zhang Ziqi, Huang Xueliang and Li Xiaohui

School of Electrical Engineering, Southeast University, Nanjing 210096, China

*tmac132180@ 163.com

Abstract. Aiming at the characteristics of Electric Vehicles (EVs) combined with transportation and mobile load, and their charging and driving behaviors will interact with the traffic system and power grid system. On the basis of this, a route planning and charging navigation strategy for EVs based on real-time traffic information and grid information was presented in this paper. First, a dynamic traffic network model was established according to the time-varying feature of traffic information. And also, the road resistance model with road segment impedance and intersection node impedance was proposed depending on the attribute of urban roads. Furthermore, based on the establishment of traffic network model, distribution network model and single EV model, a multi-objective optimization function integrated with the road travel time, the charging station load and the quantity of vehicles entering into the station was determined, which was solved to recommend the optimal driving and charging paths by dynamic Dijkstra dynamic search algorithm. At last, the actual road network of a certain zone in Nanjing was selected as an example to analyze the spatial-temporal distribution of EV charging load, and evaluated the impact of its charging and driving on traffic network and distribution network. Simulation results demonstrate its effectiveness and feasibility.

1. Introduction

With the adjustment of the world energy industry structure and people's constant attention to environmental issues, the environmental pollution and reliance on resources of fossil fuels caused by the development of automobile industry are all challenging concerns of the present world [1]. Alternative fuel vehicles, especially the plug-in electric vehicles (PEVs), are considered as an emerging solution to these aforementioned challenges, and thus providing an attractive perspective of transportation sectors and power departments [2].

However, the driving and charging characteristics of multitude EVs will have a certain impact on power grid operation and traffic congestion [3]. Therefore, an effective charging scheduling strategy is adopted to plan the optimal driving path and recommend the reasonable Charging stations (CS) for individual EV, which can reduce the owner’s range anxiety caused by battery capacity limitation, and ameliorate the influence of EV driving and charging process on power grid and traffic network[4].

As regards the charging scheduling strategy for the interaction between EVs and power grid, with the proposal of V2G (Vehicle to Grid) technology [5], various scholars have developed the orderly charging scheduling methods from the perspective of optimizing the operation of power grid and improving the interests of owners [6]. Yagcitekin et al. [7] presented a double-layer smart charging management algorithm (SCMA) for EVs based on charging safety factors. On the premise of

Content from this work may be used under the terms of the Creative Commons Attribution 3.0 licence. Any further distribution of this work must maintain attribution to the author(s) and the title of the work, journal citation and DOI.
Published under licence by IOP Publishing Ltd

satisfying both power grid and drivers' requirements comprehensively, this approach decreased the charging costs and prevented the overloading of transformers. Mo et al. [8] addressed the impact of EV charging on power quality of distribution network, and proposed an optimal EV rapid charging navigation strategy based on the internet of things network. Xu et al. [9] established a hierarchical control charging model, and analyzed the defining relationship among the different interactive systems via multiple time scales. The optimization strategy was adopted to improve the resource utilization rate of charging stations and reduce the unbalanced distribution of charging loads.

On the other hand, given the charging scheduling strategy for the interaction between EVs and traffic network, researchers mainly focus on charging navigation and path guidance for vehicles combining with traffic network information and charging infrastructure layout. In [10], a hierarchical navigation strategy based on dynamic data was proposed, considering the scenarios with traffic network information and ambient temperature. In [11], an analysis technology with big data was utilized to mine the taxi travel dataset and recommend charging stations for vehicles with charging requirements in real-time. In [12], an effective charging navigation strategy was formulated for EVs driving on the expressway, which aims to reduce the travel time and charging time cost.

In summary, although the above two types of charging scheduling strategies have achieved some results, they are only independent research from the interaction mechanism between EVs with power grid system or traffic system. However, the charging and driving behaviors of EVs are closely coupled with the power grid and traffic network, which will be affected by the interaction of power grid operation status and traffic flow information simultaneously. Accordingly, Guo et al. [13] constructed a charging path navigation architecture that integrated grid information and traffic information. It pushed the charging station capacity information to each owner through the Intelligent Transport System (ITS) and guided it to charge quickly. However, the proposed model failed to present an optimal path recommendation scheme for an EV. Furthermore, Mu et al. [14] proposed a charging load forecasting method based on the Vehicle-Road-Grid fusion model. Origin Destination (OD) method was adopted to model EVs with different functions, and the distribution network power flow analysis was carried out by integrating traffic and grid information. However, the established traffic network model did not consider the impact of impedance delay at road network intersections. Under the Vehicle-Road-Grid fusion framework, other researchers [15] optimized the EV charging path, which reduced the road congestion caused by aggregation charging and improved the space-time distribution of charging station loads. However, the traffic side model was a static road network without considering the real-time dynamic characteristics of traffic information.

Therefore, this paper proposed an EV path planning and charging navigation strategy integrating real-time traffic information and grid information. Firstly, the dynamic traffic network model, the distribution network model and the single EV model were established respectively. And also, the comprehensive road resistance target of navigation method was determined, and the multi-objective path planning problem was solved by dynamic Dijkstra algorithm. Finally, the actual road network of Nanjing was taken as an example to analyze the temporal-spatial distribution as well as the interactive influence of charging loads, and verified the implementation effect of the proposed strategy.

2. Electric vehicle - distribution network - transportation network interaction model

2.1. Dynamic traffic network model

Transportation network bears all the trip requirements, which is the basis of route planning and navigation. Due to the dynamic time-varying characteristics of traffic information, graph theory analysis method [15] was adopted to model and describe the traffic network:

$$\begin{cases} \boldsymbol{G}^{T}=(\boldsymbol{V},\boldsymbol{E},\boldsymbol{K},\boldsymbol{W}) \\ \boldsymbol{V}=\left\{v_i \middle| i=1,2,3,\mathrm{K},n\right\},\ \boldsymbol{E}=\left\{v_{ij} \middle| v_i\in\boldsymbol{V},v_j\in\boldsymbol{V},i\neq j\right\} \\ \boldsymbol{K}=\left\{k \middle| k=1,2,3,\mathrm{K},m\right\},\ \boldsymbol{W}=\left\{w_{ij}^{k} \middle| v_{ij}\in\boldsymbol{E},k\in\boldsymbol{K}\right\} \end{cases} \tag{1}$$

where, G^T denotes the traffic network; V denotes the set of all nodes in G^T, namely, the set of intersection nodes; E denotes the set of all directed arcs in G^T, namely, the set of road segments in the traffic network; W denotes the set of road segment weights; K denotes the set of time periods.

Aiming at most of the urban traffic road network nodes are equipped with signal lights. The time consumption of vehicles in the driving process is affected not only by the road impedance on the arc, but also by the time delay caused by the control of signal lights at intersections. Therefore, the road resistance consists of the road segment impedance and intersection node impedance, the urban road resistance model can be expressed as follows:

$$w_{ij}^k(t) = Cv_i(t) + Rv_{ij}(t) \tag{2}$$

where: $Cv_i(t)$ represents the intersection node impedance model, $Rv_{ij}(t)$ represents the road segment impedance model.

According to the urban traffic condition classification standard, the road traffic condition was divided into four grades by the saturation S evaluation index [16]: Unblocked ($0 < S \le 0.6$), Slow ($0.6 < S \le 0.8$), Congestion ($0.8 < S \le 1.0$) and Serious Congestion ($0.8 < S \le 1.0$). The models corresponding to different saturation are as follows:

1) Road segment impedance

$$Rv_{ij}(t) = \left\{R^1v_{ij}(t): t_0(1+\alpha(S)^\beta), 0 < S \le 1.0;\ R^2v_{ij}(t): t_0(1+\alpha(2-S)^\beta), 1.0 < S \le 2.0\right. \tag{3}$$

where: saturation $S = \dfrac{Q}{C}$: Q is the traffic flow of a road segment, and C is the traffic capacity, t_0 is the zero-flow travel time, α, β are the impact factors of the impedance model.

2) Intersection node impedance

$$Cv_i(t) = \left\{C^1v_i(t): \frac{9}{10}\left[\frac{c(1-\lambda)^2}{2(1-\lambda S)} + \frac{S^2}{2q(1-S)}\right], 0<S \le 0.6;\ C^2v_i(t): \frac{c(1-\lambda)^2}{2(1-\lambda S)} + \frac{1.5(S-0.6)}{1-S}S, \quad S > 0.6\right. \tag{4}$$

where: c is the cycle length; λ is the green split; q is the vehicle arrival rate of the road segment.

From equation (3) and equation (4), it can be known that saturation S is the only dynamic variable in the above-mentioned impedance model, and the remaining parameters are the inherent attribute values of road design and planning. Finally, according to the saturation grade of the traffic network, the intersection node impedance and road segment impedance were combined to obtain the dynamic road resistance model of the urban road network:

3) Dynamic road resistance model

$$w_{ij}^k(t) = \begin{cases} R^1v_{ij}(t) + C^1v_i(t), \quad 0<S \le 0.6;\ R^1v_{ij}(t) + C^2v_i(t), \ 0.6<S \le 0.8 \\ R^2v_{ij}(t) + C^1v_i(t), \ 0.8<S \le 1.0;\ R^2v_{ij}(t) + C^2v_i(t), \ 1.0<S \le 2.0 \end{cases} \tag{5}$$

2.2. Distribution network model

In order to match the scale of the transportation network, it is necessary to establish a certain scale and level of the distribution network. In this paper, the distribution network model based on the IEEE-33 nodes was selected, and the fast charging station was connected to the distribution network through 10/0.4 kV distribution transformer. The distribution network model is as follows:

$$D_n^P = (N_d^P, B^P, G^P) \tag{6}$$

where: N_d^P denotes parameters such as node type, power and voltage; B^P denotes parameters such as branch connection, resistance and transmission power; G^P denotes information such as location and capacity of the source node.

Further, the total charging load P_k of the distribution network node K at each moment was the cumulative power of the recharging vehicles at the charging station:

$$P_k = \sum_{i=1}^{N} P_{i,t} \tag{7}$$

where: N represents the total number of recharging vehicles at time t, and $P_{i,t}$ represents the charging power of the ith vehicle at time t.

2.3. Single Electric Vehicle model

In terms of the EV charging and driving characteristics, the modeling and analysis of single electric vehicle were carried out:

1) Initial location information

The EVs were uniformly numbered $N_e = \{i | i = 1,2,3,K .n\}$, and n represents the quantity of these introduced vehicles. OD starting and ending matrix method was adopted to simulate vehicles driving process [14], that is, the OD matrix was deduced by the change of traffic flow in each period of a road segment. For each EV introduced, its departure location $D_s(i)$, arrival location $D_d(i)$, departure time $t_s(i)$ and arrival time $t_d(i)$ were assigned via Monte Carlo Sampling [9]. OD starting and ending matrix probability matrix are as follows:

$$C_{ij}^k = \frac{Q_{ij}^k}{\sum_{j=1}^{M} Q_{ij}^k}, (1 < i < n, 1 < j < n, i \neq j) \tag{8}$$

where: b_{ij}^K denotes the traffic flow Q between initial traffic node i and destination node j in the Kth time period, and C_{ij}^k denotes the travel probability between the two nodes.

2) Battery energy information

We selected BYD E6 as the EV simulation model and referred to its technical index: battery capacity is $24\,\text{kW}\cdot\text{h}$, the endurance mileage is 160 km under operating conditions, and the power consumption per kilometer is about 0.16 (kW.h)/km. According to China's QC/T 841-2010 'Electric Vehicle Conductive Charging Interface' charging facility standard, the fast charging power was deployed as 50 kW (L3 level). In order to prevent battery damage caused by overcharging, the State of Charge (SOC) of EVs after finishing charging was set as 0.8~0.9 [13], and then the battery capacity at the initial time was [(0.8~0.9)· $B_f(i)$] kW·h. Assuming that the EV power consumption increases linearly with the travel distance [12], then the remaining battery capacity $B_t(i)$ at time t is:

$$B_t(i) = \eta(B_{t-1}(i) - \Delta l \cdot E_c) \tag{9}$$

where: η denotes the energy consumption coefficient; B_{t-1} denotes the remaining battery capacity at time t-1; Δl denotes the distance of the ith vehicle traveled from time t-1 to time t.

3) Charging demand information

When the remaining battery capacity $B_t(i)$ satisfies any condition of equation (10), the charging demand will be triggered:

$$\begin{cases} B_t(i) \leq \varepsilon \cdot B_l(i) \\ B_t(i) \leq L_{td}(i) \cdot E_c \end{cases} \tag{10}$$

where, ε is owners' range anxiety coefficient, uniform distribution ε~U[0.15,0.3] [6]. L_{td} denotes the distance from current location $D_t(i)$ to destination $D_d(i)$. That is, when the remaining battery capacity $B_t(i)$ is lower than the initial set threshold or cannot reach the destination, the charging demand is generated.

Dynamic Dijkstra path search algorithm [17] was adopted to plan the driving route for vehicles and recommend a charging station for the EV with charging demand.

$$Ne(i) = \begin{cases} j, Ch(i) = 1 \\ 0, Ch(i) = 0 \end{cases} \tag{11}$$

where: $j = 1, 2, 3, K .m$, m represents the number of charging stations, and $Ch(i)$ represents the charging demand marker. When the EV has a charging demand, it is set to 1; otherwise, it is set to 0.

4) Dynamic driving information

The time from starting position $D_s(i)$ to recommend charging station position $D_c(i)$ of a vehicle through the path search algorithm was the accumulation of road resistance **W** of actual driving paths:

$$t_a(i) = t_s(i) + \sum_{v_{ij} \in E} \left[v_{ij} \cdot w_{ij}^k(t) \right] \tag{12}$$

where: $v_{ij} = 1$ indicates that the vehicle is in the actual driving path, otherwise it is 0. The road resistance model was calculated according to equation (5) of the actual traffic information.

The waiting charging time was the difference between the charging finish moment of the vehicle *k* reached CS at the latest and the moment when the current vehicle *I* reached CS:

$$T_w(i) = t_d(k) - t_a(i) \tag{13}$$

The charging finish moment was the sum of the vehicle's arrival moment, charging waiting time and charging time:

$$t_d(i) = t_a(i) + T_w(i) + T_c(i) \tag{14}$$

The charging time $T_c(i)$ was obtained by equation (15):

$$T_c(i) = \frac{B_o(i) - B_t(i)}{\eta_c \cdot P_c} \tag{15}$$

where: η_c represents the charging efficiency, ranging from 0.8 to 0.9.

The number of vehicles in charging station $M_n(j,t)$ at time *t*, included the charging vehicles and waiting for charging vehicles:

$$M_n(j,t) = \sum Cs(j,i,t_a) \tag{16}$$

where: $Cs(j,i,t_a)$ means the *i*th vehicle staying at the recommended *j*th charging station at the time t_a .

3. Electric vehicle path planning and charging navigation method

In the interactive architecture model, in order to integrate the three information and formulate a comprehensive goal for path planning and charging navigation, we coupled the respective objective of EVs, power grid and transportation network to obtain a comprehensive road resistance function W_G .

3.1. Comprehensive road resistance function

1) EV target road resistance function

In the light of the statistical analysis results [17], travel time is the most concerned factor for owners. Therefore, the travel time $t_a(i)$ from the initial position $D_s(i)$ to charging station position $D_c(i)$ was chosen as the EV target road resistance W^E :

$$W^E = t_a(i) = t_s(i) + \sum_{v_{ij} \in \mathbf{E}} \left[v_{ij} \cdot w_{ij}^k(t) \right] \tag{17}$$

2) Power grid target road resistance function

Power grid side mainly considers the safe and economic operation of the system as its measurement index, so the charging station load P_k connected to the distribution network was chosen as the target road resistance W^P :

$$W^P = P_k = \sum_{i=1}^{N} P_{i,t} \tag{18}$$

3) Transportation network target road resistance function

To reduce the traffic segment congestion around CS caused by EV gathering and charging, the number of charging vehicles $M_n(j,t)$ at CS was chosen as its target road resistance W^T :

$$W^T = M_n(j,t) = \sum Ne(i,t_a) \tag{19}$$

Finally, the comprehensive road resistance function W^G of each single target was determined by weight coefficients:

$$W^G = \mu_1 \frac{W^E}{\min(W_S^E)} + \mu_2 \frac{W^P}{\min(W_S^P)} + \mu_3 \frac{W^T}{\min(W_S^T)} \tag{20}$$

where: μ_1, μ_2 and μ_3 respectively represent the weight coefficients, W_S^E, W_S^P and W_S^T independently optimize the results of travel time, charging station load and the number of vehicles at charging station without considering the other two objectives. It can be seen that the optimal charging strategy recommends the charging station with short travel time, low charging load and few incoming vehicles.

3.2. Constraint conditions

To ensure the rationality of charging navigation strategy, the constraints are given as follows:

1) EV charging constraints

The mileage of recommended charging stations for EVs with charging demand should be within the range of remaining electricity:

$$B_t(i) \le L_{tc}(i) \cdot E_c \tag{21}$$

where: $L_{tc}(i)$ represents the distance from the vehicle's current position $D_t(i)$ to CS position $D_c(i)$.

2) Grid operation constraints

The charging station load at each moment should be less than the set threshold:

$$P_k(j,t) < P_{thr}(j,t) \tag{22}$$

where: $P_k(j,t)$ denotes the load of the *j*th charging station at time *t*; $P_{thr}(j,t)$ denotes the load threshold of the *j*th charging station at time *t*.

Similarly, to ensure the safe operation of the distribution network, the voltage offset rate of each node should be within a reasonable range:

$$|N_s(k,t)| < 7\% \tag{23}$$

where: $N_s(k,t)$ represents the voltage offset rate of the *k*th node at time *t* in the distribution network.

3) Traffic road travel constraints

To avoid the traffic jam, the driving speed of the recommended road segment should be restricted:

$$\left\{ V_e(i,t) \ge 30\% V_0;\ V_e(i,t) = \frac{L_{ij}}{Rv_{ij}(t)};\ V_0 = \frac{L_{ij}}{t_0} \right. \tag{24}$$

where: $V_e(i,t)$ represents the average travel speed of actual segment, V_0 represents the average travel speed at zero-flow, and L_{ij} represents the length of the road segment v_{ij}.

4. Path planning and charging navigation flowchart

Figure 1 shows the EV path planning and navigation flowchart, the whole process is as follows:

Update of information parameters: Firstly, the system initializes the real-time information of EVs, power grid and traffic network, and reads the initial parameters of EV position and battery energy.

Driving path planning: The system judges the charging demand marker, and when the charging demand is not triggered, the goal is to drive with the minimum road resistance $w_{ij}^k(t)$, that is, the minimum travel time-consuming. Dynamic Dijkstra algorithm is utilized to calculate road resistance information, and three driving routes are recommended for the owner, who randomly chooses any route to drive. Current position $D_t(i)$ and speed of each vehicle $V_e(i)$ are recorded, the remaining battery capacity $B_t(i)$ is updated after reaching the destination, and the travel path planning is completed.

Charging path navigation: Once the charging demand triggers, the minimum comprehensive road resistance is taken as the charging navigation target. The road segment is assigned by calculation W^G, and the dynamic Dijkstra algorithm is adapted to search and recommend three charging paths. The owner randomly chooses any route to drive for recharging. When an EV arrives at a station, the number of vehicles at the a station and their arrival time $t_a(i)$ are counted. Vehicles are charged with constant power P_c. Determine whether the charging is completed or not, and if it is completed, change the travel position to destination position $D_d(i)$ again, so that the travel route navigation planning can be carried out with the least travel time-consuming until the destination is reached.

System scheduling judgment: All vehicles introduced are judged by charging schedule. When all those vehicles complete path planning and charging navigation, they exit the scheduling.

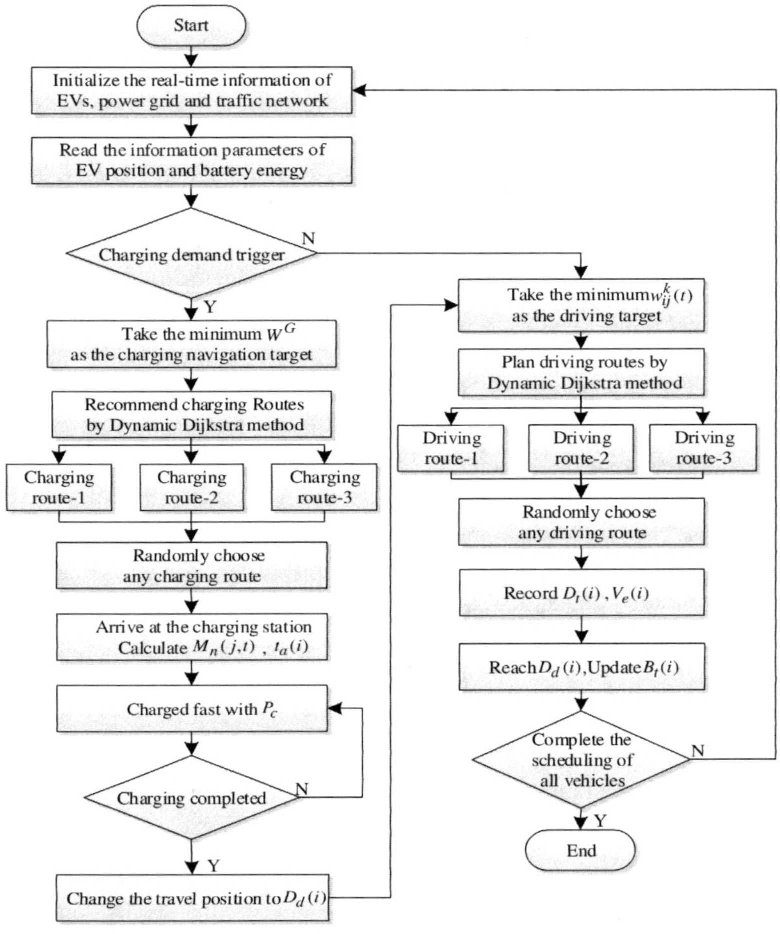

Figure 1. Flowchart of EV path planning and charging navigation.

5. Case studies and analysis

5.1. Actual road network model

To verify the implementation effect of the charging navigation strategy proposed, the actual road network was selected for modeling and analysis. It was divided into residential areas(RA), commercial areas(CA) and industrial areas(IA) according to the standard of the traffic function area. Figure 2 shows the road network topology diagram of a certain area in Nanjing. As depicted, the total area is about 48.8 km^2, and the circumference is about 27.9 km, including 103 intersection nodes and 158 major road segments. The model parameters and real-time traffic information can refer to our previous research in [18].

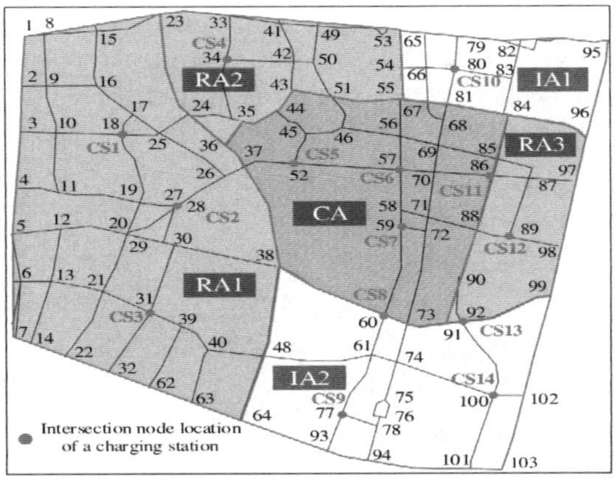

Figure 2. Topological structure of a regional road in Nanjing.

Depending on the 'Implementation Plan for Promotion and Application of New Energy Vehicle in Nanjing during the 13th Five-Year Plan', the specific number of vehicles introduced in each region is as follows: RA1(7160 vehicles), RA2(4296 vehicles), RA3(2864 vehicles), IA1(7432 vehicles), IA2(1856 vehicles) and CA(7160 vehicles). And also, in order to match the scale of the above-mentioned traffic network, we established three distribution network models of IEEE-33 nodes. The charging stations are connected to the distribution network through nodes, whose load threshold is 2 MW and the prescribed capacity is 30 vehicles.

5.2. Analysis of simulation results

Firstly, we started from the overall optimization of the system and did not study the influence of weight coefficient on the results in depth. Thus, the weight coefficients were uniformly set as 1/3. Figure 3(a) and Figure 3(b) show the spatial and temporal distribution of all-day charging demands.

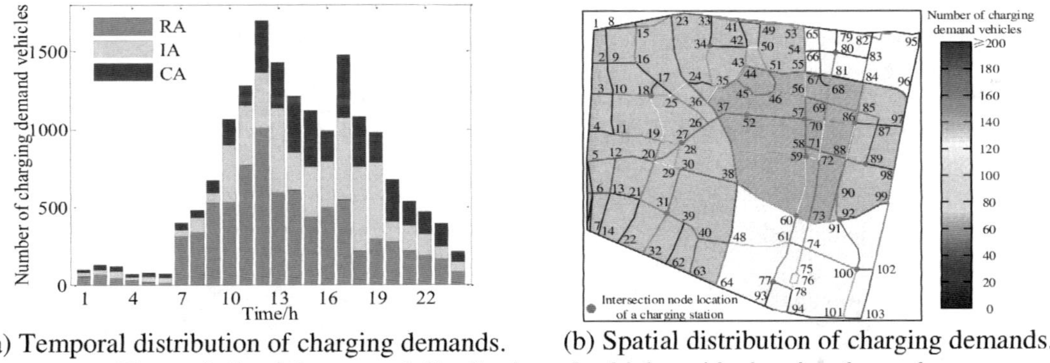

(a) Temporal distribution of charging demands. (b) Spatial distribution of charging demands.

Figure 3. Spatial-temporal distribution of vehicles with charging demands.

As revealed by Figure 3, on the whole, the temporal distribution of charging demands is not uneven. Their peak time periods occur in 11:00-12:00 and 16:00-17:00. Residential peak concentrates in 11-12:00 and 16:00-17:00, industrial peak concentrates in 12:00-13:00 and 17:00-18:00, and commercial peak concentrates in 13:00-14:00 and 17:00-18:00. The temporal distribution of charging demand is consistent with the rule distribution of residents' trips.

In terms of spatial distribution, there are some differences according to the types of functional areas. The number of vehicles needed for recharging in commercial areas is the largest, with an average number of 45.56 vehicles, followed by 32.15 vehicles in residential areas, and the least in industrial areas, with an average number of 19.42 vehicles.

5.2.1. Analysis of single EV charging navigation path. Based on the charging demand distribution, to describe the EV charging navigation path, the EV numbered $Ne(12587)$ was selected for path analysis. Its OD positions are node 9 and node 74, travel time is at 08:15. Meanwhile, the driving path of disorderly charging method is compared with our optimal driving path which is planned according to the goal of the shortest distance for vehicles. The nearest charging station is recommended for an owner when the charging demand is triggered. The analysis results are listed in Table 1.

Table 1. Results of disorderly charging path and optimal charging path

Method	Actual route	Recommended CS	Driving mileage/km	Travel time/min	Charging waiting time/min	Charging station load/MW	Number of vehicles at station
Disorderly route	9→10→11→19→20 →28→30→31→39 →40→48→61→74	CS2	6.2	34.55	8.8	1.75	48
Optimal route-1	9→10→18→19→20 →29→31→39→40 →48→61→74	CS3	9.1	21.78	4.1	1.30	24
Optimal route-2	9→10→18→25→26 →37→52→37→38 →48→61→74	CS5	10.8	25.34	6.8	1.15	18
Optimal route-3	9→16→17→19→27 →28→29→31→39 →40→48→61→74	CS3	12.7	27.88	4.6	1.35	27

As depicted, although the disorderly charging method achieves the least driving mileage, compared with the optimal charging method, it does not avoid congested roads and takes a long time, which exceeds 36 % of that average value. And CS2 is recommended nearby for recharging, whose charging load is close to the limit threshold, and the number of vehicles at the station exceeds the limit capacity by 18. However, the three recommended paths by the optimal method have less travel time and average waiting time of 5.1 minutes. Although it is in the morning peak period, the indexes tend to be saturated, but they are within the controllable capacity range.

5.2.2. Analysis of power grid impact. And then, the influence of charging behaviors on the distribution network was analyzed. Load distribution of each charging station under disorderly method and optimal method are shown in Figure 4(a) and Figure 4(b).

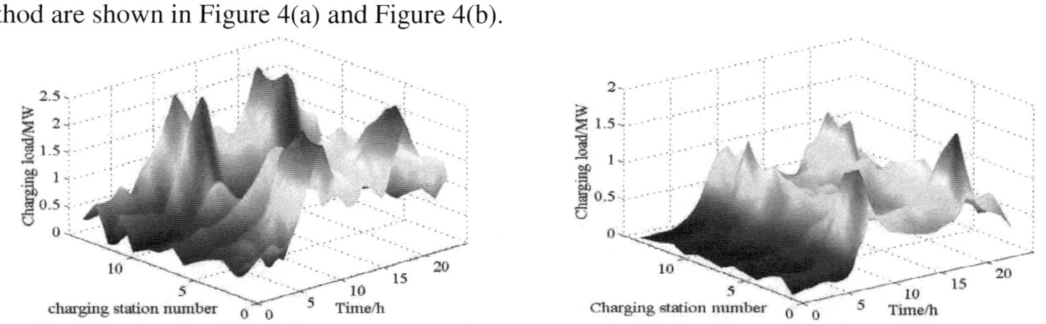

(a) Charging station load of disorderly method. (b) Charging station load of optimal method.

Figure 4. Charging station load obtained by disorderly method and optimal method.

From Figure 4(a) and 4 Figure(b), under the disorderly charging strategy, the distribution of loads is uneven and the peak-valley difference of each station is large. Among them, the maximum load is CS11 with 2.45 MW at 09:30, which is beyond the limit threshold, the minimum load is CS4 with 0.20 MW at this time. Meanwhile, under the optimal charging method, the CS11 load is reduced to 1.55 MW and the lowest CS4 load is increased to 0.35 MW, which reduces the load peak-valley difference.

5.2.3. Analysis of traffic network impact. At last, the traffic travel capacity of the road segment was evaluated. Figure 5 shows the change of the average traffic speed of road segment [57-58] near the

CS5. It can be seen that the speed of vehicles in the road segment [57-58] is low between 07:00-10:00 and 18:00-21:00. Under the disorderly strategy, the proportion of time range below the zero-flow average speed is 47.8 %, and the optimal strategy is 27.6 %. Therefore, the optimal strategy by traffic side optimization improves road trip efficiency.

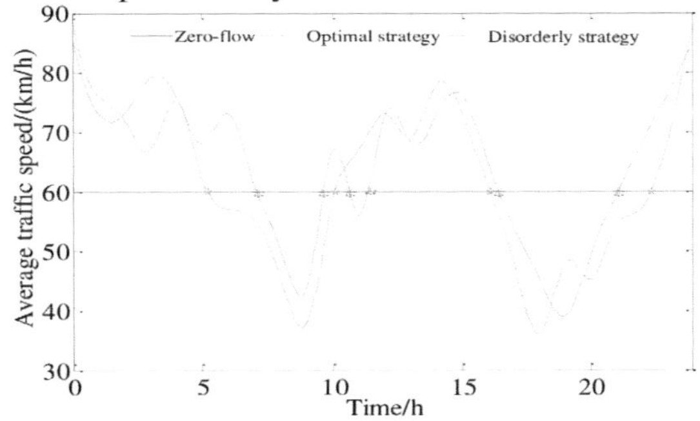

Figure 5. Average speed of vehicles in the road segment [57-58].

6. Conclusions

This paper proposed an EV path planning and charging navigation strategy, and established a dynamic traffic network model, distribution network model and single EV model respectively. Finally, taking a specified regional road network and the distribution network as examples, the validity of this strategy was verified by simulation tests. The most important results of the proposed approach are as follows.

1) The proposed interactive model and navigation strategy recommends the optimal path based on real-time road information, which can be used as a navigation algorithm to plan the driving path, and also can reasonably recommend charging station through integrated road resistance search.

2) The charging navigation strategy combining real-time traffic information and grid information, provides three driving and charging routes simultaneously, effectively diverts traffic flow, and avoids road congestion, as well as improves traffic system operating condition.

3) The optimal strategy reasonably controls the number of charging EVs and effectively dispatches the load resources of charging stations. It not only shortens the waiting time for recharging, but also improves the economy and security of charging stations.

References

[1] Du J, Ouyang M and Chen J 2017 *J. Energy.* **120** 584-96
[2] Ji Z and Huang, X 2018 *J. Renew. Sust. Energ. Rev.* **90** 710-727.
[3] Tang D and Wang P 2016 *J. IEEE Trans. Power Syst.* **31(6)** 4231-4242.
[4] Wang G, Xu Z, Wen F and Wong K P 2013 *J. IEEE Trans. Power Deliv.* **28(4)** 2363-2372.
[5] Kempton W and Letendre S E 1997 *J. Res. Part D-Transport. Environ.* **2(3)** 157-175.
[6] Habib S, Kamran M and Rashid U 2015 *J. J. Power Sources.* **277** 205-214.
[7] Yagcitekin B and Uzunoglu M 2016 *J. Appl energ.* **167** 407-419.
[8] Mo W, Yang C, Chen X, Lin K and Duan S 2019 *J. Energies.* **12(6)** 962.
[9] Xu Z, Su W, Hu Z, Song Y and Zhang H 2015 *J. IEEE Trans. Smart Grid.* **7(1)** 428-438.
[10] Xia F, Chen H, Chen L and Qin X 2019 *J. IEEE Access.* **157** 417-429.
[11] Tian Z, Jung T, Wang Y, Zhang F, Tu L, Xu C and Li X Y 2016 *J. IEEE Trans. Intell. Transp. Syst.* **17(11)** 3098-3109.
[12] Yang S N, Cheng W S, Hsu Y C, Gan C H and Lin Y B 2013 *J. Math. and Comput. Model.* **57(11-12)** 2873-82.
[13] Guo Q, Xin S, Sun H, Li Z and Zhang B 2014 *J. IEEE Trans. Smart Grid.* **5(4)** 1969-79.
[14] Mu Y, Wu J, Jenkins N, Jia H and Wang C 2014 *J. Appl. Energy.* **114** 456-465.
[15] Luo Y, Zhu T, Wan S, Zhang S and Li K 2016 *J. Energy.* **97** 359-368.
[16] Yu T and Ma J 2016 *In 2016 IEEE International Conf. on Intelligent Transportation Engineering.*(Singapore:IEEE) p 123-127.
[17] Mariyasagayam M N and Kobayashi Y 2013 *J. Energy.* **57** 134-142.

IOP Conf. Series: Materials Science and Engineering **752** (2020) 012011 doi:10.1088/1757-899X/752/1/012011

[18] Li X H, and Li L 2019 *J. Power System Protection and Control*. **42** 140-149.

IOP Conf. Series: Materials Science and Engineering **752** (2020) 012012 doi:10.1088/1757-899X/752/1/012012

Research on Charging Load Characteristics of EVs Based on Actual Charging Power

Min Wang[1], Peng Zhang[1,*], Yi Shi[1], Chao Wu[1], Mahammad Mammadli[1] and Wei Zhang[2]

[1] College of Energy and Electrical Engineering, Hohai University, 8 west focheng road, Nanjing 211100, Jiangsu, China
[2] NARI Group Corporation (State Grid Electric Power Research Institute), Nanjing 211100, Jiangsu, China

[*]531689065@qq.com

Abstract. With the expansion of the scale of electric vehicles (EVs) and the maturity of fast charging technology and the popularization of fast charging piles, the disorderly fast charging of EVs caused a significant damage on distribution network, the actual charging power of EVs is different from the rated power, the traditional method of using the rated power of the charging pile to forecast the charging load will cause a significant error and affect the dispatching operation of power grid. Based on the actual EV charging curve, this paper establishes a variable power fast charging load model for EVs. The Monte Carlo simulation method is used to simulate the charging load of EV groups. The fast charging loads of EVs with different vehicle types and different battery capacities are compared. The actual variable power fast charging mode of the electric vehicles is compared with the constant power fast charging mode and the slow charging mode, and the charging loads of the three charging modes are significantly different. Finally, the impact of charging load of electric vehicles of different scales on distribution network load is analyzed.

1. Introduction

With the rapid development of the global economy, the problem of global warming and shortage of fossil energy is becoming more and more dangerous. The popularisation of traditional fuel vehicles has been unable to meet the requirements of environmental protection, energy conservation and emission reduction, so it is urgent to develop new energy vehicles. At present, governments around the world have formulated development strategies and introduced new policies to promote the research and development and marketing of EVs [1]. Relevant enterprises actively invest and EV technology rapidly developed.

By the end of 2018, the number of EVs in the world reached 5.4 million, an increase of 64% over 2017. In the whole year of 2018, the global sales of EVs reached 2.01 million, and the share of EVs in the global automotive market reached 2.1%, an increase of 72% [2].

The charging load of EV have a randomness in time and space. The access of large-scale EVs may lead to problems such as increased peak load, partial overload, grid congestion, increased fluctuation and increased investment in distribution networks. It will have a significant impact on the stability of the power grid and the planning and operation of the power system and the operation of the power

Content from this work may be used under the terms of the Creative Commons Attribution 3.0 licence. Any further distribution of this work must maintain attribution to the author(s) and the title of the work, journal citation and DOI.
Published under licence by IOP Publishing Ltd

market. On the other hand, EVs can be scheduled and controlled as schedulable loads and energy storage, which can reduce the pressure and improve the reliability of the power grid [3-4].

Therefore, the forecasting and simulation of the charging load of EVs is very imperative. In [5], the existing charging load calculation method is divided into the determination analysis method based on travel demand, the Monte Carlo simulation method and the probability analysis method of charging station charging load, then the three methods were introduced respectively. In [6-8], they use driving pattern data from the National Household Travel Survey (NHTS 2001) to fit the probability density distributions of travel time and daily mileage of household EVs. Then, based on these probability distributions, Monte Carlo simulation is carried out to analyze the charging load of EVs. literature [9-11] subdivides the types of EVs into buses, taxis, private cars and official vehicles. According to the driving patterns of different types of vehicles, and considering different charging modes such as fast charging, conventional charging and slow charging, the charging load forecasting method of EVs is proposed. In [9], charging time is divided into three types based on the analysis of user behavior, and each factor is obtained by Monte Carlo simulation method. Literature [10,11] proposed several kinds of EV orderly charging control strategies based on peak-to-valley electricity price and considering various user response, which significantly reduced the peak load and peak-to-valley difference of the power grid.

At present, the forecasting of the charging load of EVs is mostly based on the conventional or slow charging mode, and the charging power is assumed to be constant power or evenly distributed within a specific range. Literature [12] proposed that the charging power of EVs changes with time, but its charging power model is only the product of voltage and current, and does not consider the difference between vehicle type and battery capacity, and does not compare the difference of charging loads of EVs with different speeds and the difference of constant power charging and variable power charging. Due to the maturity of EV fast charging technology and the popularization of fast charging piles, EV owners are more willing to adopt fast charging with short time and high efficiency. The disorderly fast charge of the EV will have a considerable impact on the distribution network. Therefore, the accurate simulation and forecasting the real-time charging power of EV, especially in the case of fast charging mode, is of considerable significance to distribution network dispatching and control strategy.

This paper focuses on the fast charging load characteristics of EVs based on actual charging power. Using the data of the charging power of the actual measured new mainstream type of electric vehicles which changes with the battery state-of-charge (SOC), the probability simulation of EVs can be done through Monte Carlo method which extract a single vehicle's start charging time and mileage. The actual power fast charging loads of EVs with different vehicle types and different battery capacities are compared. And the difference of EVs charging loads with different charging speeds and different scales are compared. The results show that the fast charging load curves are steep more prominent, have more significant peak value and more significant impact on the grid than slow charging load curves; in the case of a small number of EVs, the fast charging mode based on the actual charging curve will not only increase the peak value of the load curve, but also lead to increased load curve fluctuations, and it also have a significant impact on the distribution network scheduling and operation.

2. Factors affecting the charging load of electric vehicles

The charging load of electric vehicles has a high space-time randomness. And different charging load curves will be generated by considering different affecting factors. At present, the following factors affecting the charging load of EVs are mainly considered [13],

- Battery characteristics: Battery capacity, charging and discharging rate, power consumption per 100 kilometres determine the charging time and charging frequency of EVs.
- User behaviour: The driving habits and travel patterns of vehicle owners are another factor that directly affect the distribution of charging time and space and the charging load, while other factors such as weather, temperature, date type and bus scheduling affect EV charging load by influencing user behavior characteristics [14]. At the same time, buses, taxis, private cars, official vehicles and other vehicles of different uses have different charging demands.

IOP Conf. Series: Materials Science and Engineering **752** (2020) 012012 doi:10.1088/1757-899X/752/1/012012

- Charging facilities: Power replenishment modes of EVs can be divided into charging mode and electric changing mode, and charging mode can be divided into fast charging, conventional charging and slow charging. The distribution and quantity of charging facilities will affect the spatial and temporal distribution of the charging loads of EVs through users' charging demands and behaviours. Meanwhile, with the proposal and promotion of vehicle-to-grid (V2G) technology, charging facilities with two-way energy flow will become the mainstream [5]. Currently, the Alternating Current (AC) standard is adopted for conventional or slow charging facilities, while the Combined Charging System (CCS) and the Direct Current (DC) standards are adopted for fast charging facilities.
- The scale of EVs: The overall number of EVs reflects the direction of national policies and people's living standards and the development of the market economy.
- Charging control method: This includes policy regulation and market guidance, such as peak and valley electricity price and grid incentive, and regulating the temporal and spatial distribution of charging load of EVs through some charging control strategies.

This paper mainly considers the first four factors. In order to facilitate the establishment of EVs charging load model, this paper gives the following reasonable assumptions:

- The initial charging time and daily driving mileage of each type of EVs are independent variables.
- charging end marks of all EVs are SOC=1.
- The last return time of all EVs is the initial charging time.

3. Electric vehicle charging power modeling

3.1. EVs starting charging time and daily driving mileage probability modeling

The travel habits and behaviors of private vehicles owners determine the starting charging time and daily driving mileage of EVs. According to the NHTS 2001, 14% of household vehicles are not be used in one day, and 43.5% of the vehicles have a daily mileage of 20 miles (about 32 km), 83.7% of the vehicles have a daily mileage of 60 miles (about 97 km). After the normalisation of the statistical data, the return time of the final journey and daily driving mileage of the vehicles are fitted as normal distribution and lognormal distribution respectively by the method of maximum likelihood estimation. The results are shown in figure 1, figure 2 and equation 1, equation 2.

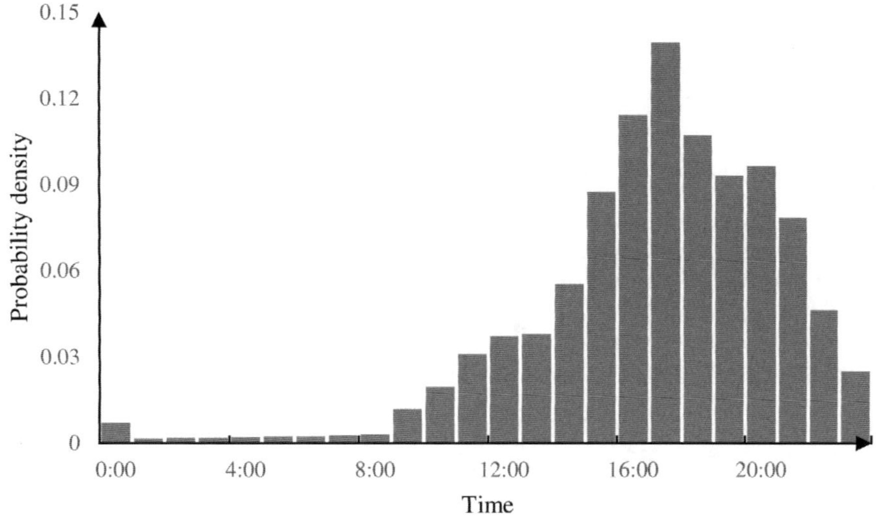

Figure 1. Probability distribution of last return time of private vehicles.

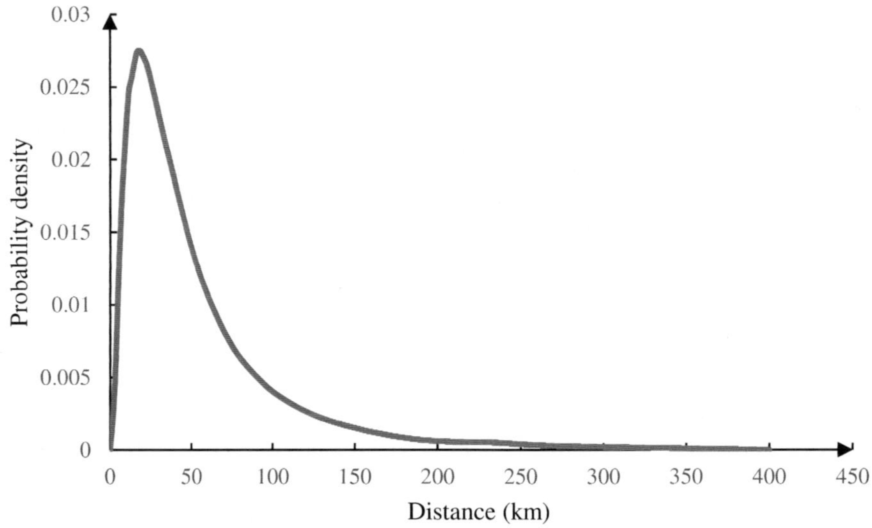

Figure 2. Probability distribution of daily mileage of private vehicles.

By comparing the fitting results in figure 1, it can be concluded that the starting charging time of private EVs satisfies the following normal distribution that the expectation μ_s = 17.6; standard deviation σ_s =3.4, and its probability density function is (1):

$$f_s = \begin{cases} \frac{1}{\sigma_s\sqrt{2\pi}}\exp[\frac{(x-\mu_s)^2}{2\sigma_s^2}], & (\mu_s - 12) < x \le 24 \\ \frac{1}{\sigma_s\sqrt{2\pi}}\exp[\frac{(x+24-\mu_s)^2}{2\sigma_s^2}], & 0 < x \le (\mu_s - 12) \end{cases} \tag{1}$$

In equation (1), x is the starting charging time.

By comparing the fitting results in figure 2, it can be concluded that the daily driving mileage of private EVs satisfies the following lognormal distribution that the expectation μ_D = 3.2; standard deviation σ_D =0.88, and its probability density function is (2):

$$f_D = \frac{1}{y\sigma_D\sqrt{2\pi}}\exp[-\frac{(\ln y - \mu_D)^2}{2\sigma_D^2}] \tag{2}$$

3.2. EVs fast charging power modeling

In this paper, based on the test data of FASTEND website [15], the fast charging power curves of three typical EV models (Kia e-Niro, BMW i3, Nissan Leaf) are selected. Charging power curves are shown in figure 3.

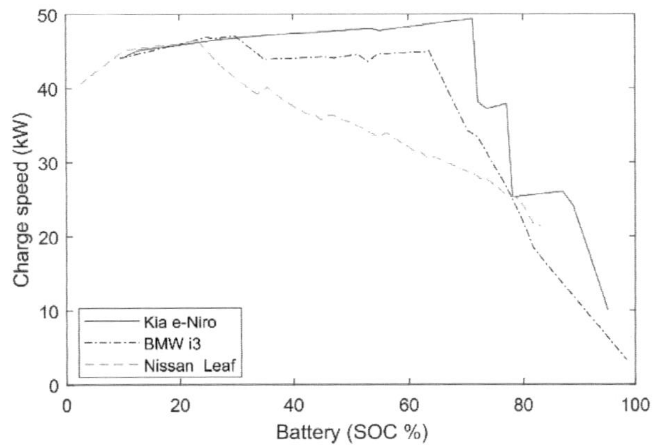

Figure 3. Charging power curve of Kia e-Niro, BMW i3 and Nissan Leaf.

According to figure 3, the charging powers of the three types of EVs are approximated fitted as follows:

$$P_1(\mathrm{SOC}_1)=\begin{cases}46 & (0<\mathrm{SOC}_1<0.7)\\ 37 & (0.7\le \mathrm{SOC}_1<0.76)\\ 25 & (0.76\le \mathrm{SOC}_1<0.86)\\ -203.4SOC_1+203.8 & (0.86\le \mathrm{SOC}_1<1)\end{cases} \tag{3}$$

$$P_2(\mathrm{SOC}_2)=\begin{cases}-166.3SOC_2^2+75.97SOC_2+38.17 & (0\le \mathrm{SOC}_2<0.35)\\ 44 & (0.35\le \mathrm{SOC}_2<0.64)\\ -119.3SOC_2+118.6 & (0.64\le \mathrm{SOC}_2<1)\end{cases} \tag{4}$$

$$P_3(\mathrm{SOC}_3)=35.9\exp(-(\frac{\mathrm{SOC}_3-0.08845}{0.2933})^2)+30.79\exp(-(\frac{\mathrm{SOC}_3-0.5758}{0.4345})^2) \tag{5}$$

In equation (3)-(5), $P_1(\mathrm{SOC}_1)$ is the charging power of Kia e-Niro with 64kWh lithium polymer battery under 50kW charging pile; $P_2(\mathrm{SOC}_2)$ is the charging power of BMW i3 with 22kWh lithium ion battery under 50kW charging pile; $P_3(\mathrm{SOC}_3)$ is the charging power of Nissan Leaf with 24kWh lithium battery under 50kW charging pile; SOC_1, SOC_2, SOC_3 is the real-time charging state of three types of vehicle batteries.

3.3. EVs state-of-charge calculation modeling

$$SOC_i(t)=SOC_i(t-1)+P_i(SOC_i(t-1))\Delta t/60/C_i \tag{6}$$

In equation (6), $SOC_i(t)$ is the charging state of the *i*-th EV battery at time *t*; $P_i(SOC(i))$ is the charging power of EVs, which is obtained by the equation (3)-(5); Δt is the unit charging time set in the simulation and its value is 1min; C_i is the battery capacity of the *i*-th EV.

3.4. Fast charging time modeling of EVs under actual charging curve

EV charging time model in general constant power charging mode is as follows:

$$T_c = \frac{DW}{100P_c} \tag{7}$$

In equation (7), T_c is the length of charging time; D is the daily driving mileage; W is the power consumption of 100 kilometres; P_c is the charging power of a EV.

Due to the change of charging power with SOC under the variable power charging mode, the above charging time calculation equation is no longer applicable. The total charging time length T_{cc} in variable power charging mode needs to be solved by the following equations:

$$\frac{DW}{100} = \int_0^{T_{cc}} P(SOC(t))dt \tag{8}$$

$$T_{cc} = T \quad if \sum_{t=1}^{T} P(SOC(t)) = \frac{DW}{100} \tag{9}$$

Equation (8) is discretized and becomes equation (9). Here T is the time length of EV that has been charged.

3.5. Fluctuation index

In order to show the fluctuation of the load, a fluctuation index is proposed here, defined as follows:

$$S = \frac{\max \sum_{i=1}^{m} (maL(i) - miL(i))}{\max(L)} \tag{10}$$

In equation (10), L is the charging load or grid load; maL and miL are local maxima and minima of L; m is the quantity of the maxima or minima.

4. Charging load simulation of EV group based on Monte Carlo method

The charging load of the EV group samples at a specific time is the sum of the charging powers of all EVs at that time. The equation is as follows:

$$P_\Sigma = \sum_{j=1}^{N_1} P_{1\cdot j} + \sum_{k=1}^{N_2} P_{2\cdot k} + \sum_{l=1}^{N_3} P_{3\cdot l} \tag{11}$$

In equation (11), N_1, N_2, N_3 are the number of samples for the three EV groups respectively.

According to the EVs charging load simulation model in section 3, the Monte Carlo method is used to extract the initial charging time and the daily driving mileage of each samples. The simulation steps are shown in figure 4:

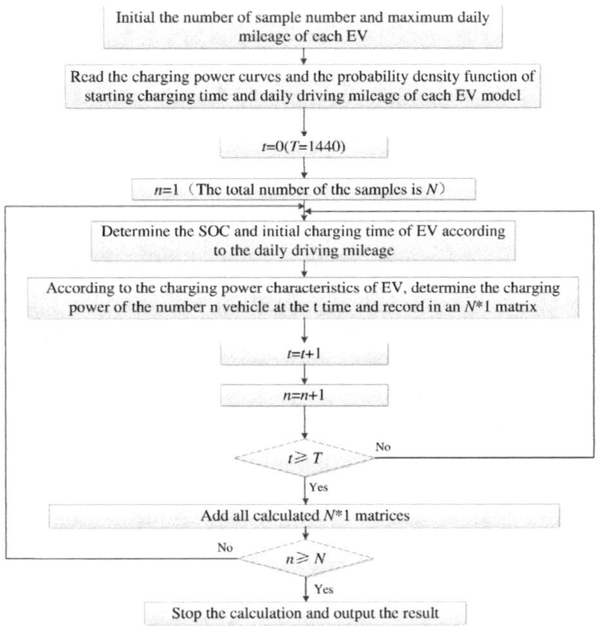

Figure 4. Simulation flowchart of charging load of EV group based on Monte Carlo method.

5. Simulation results and analysis

In this section, three types of EVs (Kia e-Niro, BMW i3 and Nissan Leaf) mentioned in subsection 3.2 with typical fast charging power characteristics are the research objects. The maximum mileages of three EVs is 485km, 240km, 250km, and battery capacities are 64 kWh, 22 kWh and 24 kWh. The difference between the actual fast charging loads of EV groups of different types is compared. And the charging loads difference between different scales and different charging schemes of EV groups are studied. At the same time, taking the typical daily load curve of Shanghai in 2018 as an example, the impact of EVs of different scales on the load curve of the distribution network is studied.

5.1. Comparison of fast charging loads of EVs in three types

In this example, the EVs are divided into three groups, each of which has 10,000 vehicles. The EVs in the same group have the same type, namely the Kia e-Niro, the BMW i3 and the Nissan Leaf. The charging loads of the three EV groups in one day are simulated as follows:

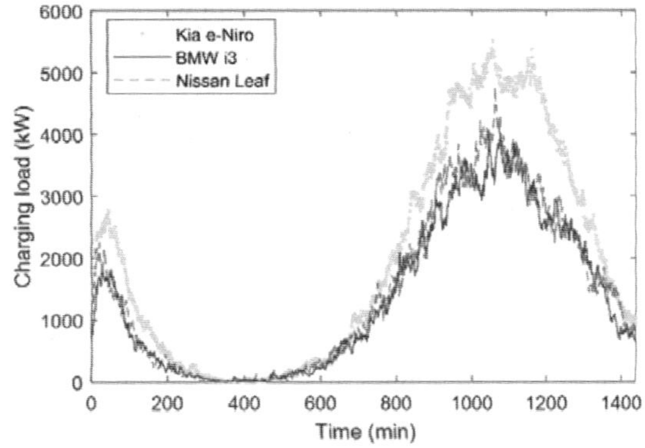

Figure 5. Three types of EV group daily charging loads.

Figure 5 shows that the three charging power curves have significant differences due to the difference in battery capacity and maximum mileage of the three types of EVs. From the battery capacity and maximum mileage of EVs, the power consumption per 100km of Kia e-Niro, BMW i3 and Nissan Leaf can be roughly calculated to 13.34 kWh, 9.17 kWh and 9.6kWh respectively, and the ratio of the three corresponds to the curve height ratio. Therefore, it will cause significant errors without considering the actual EV type and battery difference, and the simulation of this example is more practical for actual grid scheduling and load forecasting.

5.2. Analysis of charging load characteristics of EV Groups under different charging schemes

In order to compare the impact of the charging load of the EVs on the grid load under different charging modes, the three types of EVs are selected in the 1:1:1 mix, and this example calculate the charging loads when using the following three charging modes for 3,000, 30,000, and 300,000 EVs respectively:

(1) Fast charging according to the actual charging curves under the rated 50kW fast charging piles;

(2) Constant speed fast charging with a rated power of 50 kW;

(3) Slow charging with a uniform distributed power between 2 kW and 3 kW.

The charging load daily variation curves of the three modes are shown in figure 6-8.

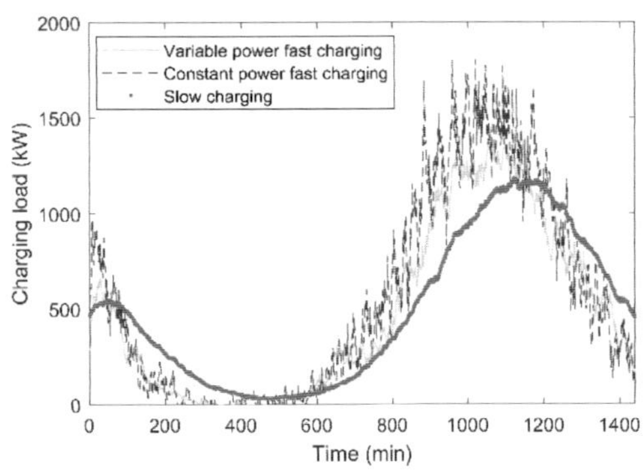

Figure 6. Charging load of 3,000 EVs of three charging modes in one day.

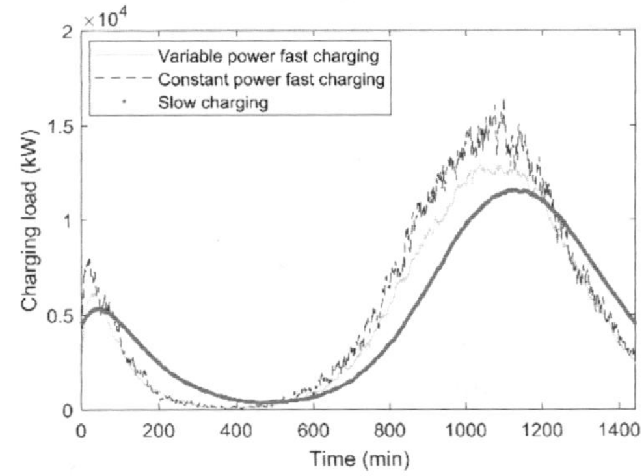

Figure 7. Charging power of 30,000 EVs of three charging modes in one day.

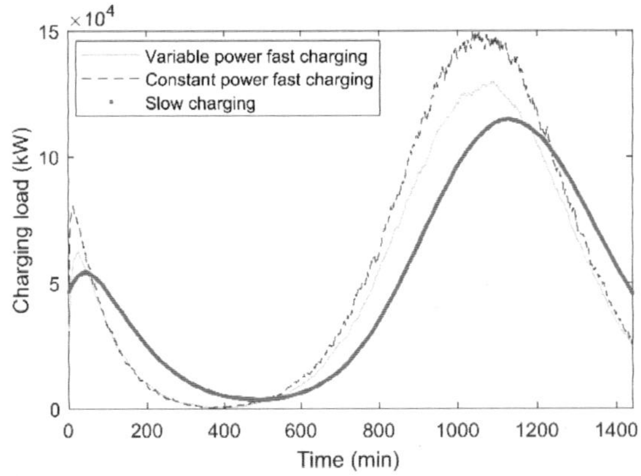

Figure 8. Charging power of 300,000 EVs of three charging modes in one day.

The results show that:

(1) When the EVs are in slow charging mode with a power uniform distributed between 2 and 3 kW, the charging load curve is gentler, the power fluctuation is smaller, the peak value of the curve is smaller, and the time of occurrence of the peak value is later than that of the fast charging mode.

(2) In the two fast charging modes, the peak-to-valley difference of the charging curve is more considerable, and the short-term power fluctuation is relatively large, especially in the case where the number of EVs is small. figure 6 shows that when 3,000 EVs are charging, the maximum fluctuation value is 261kW and the fluctuation index $S = 18\%$ in mode (1), and the maximum fluctuation value is 648kW and the fluctuation index $S = 35\%$ in mode (2).

(3) There are also significant differences in the charging power curve and the maximum charging power in the two fast charging modes. For example, when only 3,000 vehicles are charging, the maximum power difference between the two fast charging modes is 592 kW. When 30,000 vehicles are charging, the maximum power difference between the two charging modes is 2435 kW. When 300,000 vehicles are charging, the maximum charging power of the two fast charging modes differs by 21651 kW. It can be seen that if the constant power charging mode is adopted in the distribution network operation and scheduling to do the simulation and forecasting of the charging power will cause a significant error with the actual value.

5.3. The impact of EVs fast charging load on the distribution network

By the end of 2018, the number of motor vehicles in Shanghai had exceeded 3 million. It is estimated that the increasing number of EVs is expected to exceed 30% in the next five years. Therefore, this example uses the typical daily load data of Shanghai in the summer of 2018. The three types of EVs are combined in a 1:1:1 mix. A total of 100,000 and 300,000 EVs are added to analysis the impact to the peak load in Shanghai when charging in variable power disordered fast charging mode. The original load and total load during peak load time are shown in figure 9. It shows that after adding 300,000 EVs, the peak value of the distribution network load curve is increased by 70MW, and the difference is significant. Therefore, it is very important to carry out accurate simulations for specific models and specific charging power characteristics in the study.

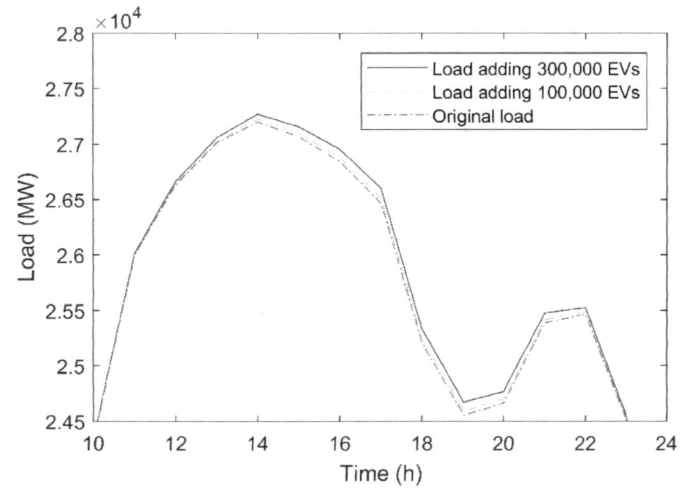

Figure 9. Partial diagram of the typical daily load curve of Shanghai after adding EVs.

6. Conclusion

Based on the actual variable power fast charging characteristics of multi-type EVs, this paper extracts the initial charging time and daily driving mileage of EVs through Monte Carlo simulation method to simulate the actual variable power fast charging load of EVs. The fast charging loads of EVs with different vehicle types and different battery capacities are compared. Moreover, the actual variable power fast charging mode is compared with the rated constant power fast charging and slow charging modes. The results show that when the rated charging power of the EV charging pile is constant, if the charging load of the hybrid EV group is calculated only by the rated power, a significant error will occur. At the same time, the actual variable power fast charging mode has a more significant impact on the grid than the conventional or slow charging mode, and the actual variable power fast charging load variation law of different EV types are different. Therefore, using the actual power fast charging characteristic curve of different EV types to simulate the charging load of the EV group will improve the accuracy and practicability of various research and analysis.

References

[1] Hafez O and Bhattacharya K 2017 Queuing analysis based PEV load modeling considering battery charging behavior and their impact on distribution system operation *IEEE Trans. Smart Grid* **9** 261
[2] Roland I and EV-volumes.com 2018 *available: http://www.ev-volumes.com/country/total-world-plug-in-vehicle-volumes*
[3] Clairand J, Rodríguez-García J andÁlvarez-Bel C 2018 Smart charging for electric vehicle aggregators considering users' preferences *IEEE Access* **6** 54624-35
[4] Bashash S and Fathy HK 2012 Transport-based load modeling and sliding mode control of plug-in electric vehicles for robust renewable power tracking *IEEE Trans. Smart Grid* **3** 526-34
[5] Wang XF, Shao CC, Wang XL and Du C 2013 Survey of electric vehicle charging load and dispatch control strategies *Proceedings of the CSEE* **33** 1-10
[6] Tian LT, Shi SL and Jia Z 2010 A statistical model for charging power demand of electric vehicles Power System Technology *Power System Technology* **34** 126-30
[7] Taylor J, Maitra J, Alexander M, Brooks D and Duvall M 2009 Evaluation of the impact of plug-in electric vehicle loading on distribution system operations *2009 IEEE Power & Energy Society General Meeting* (Calgary) pp 1-6
[8] Shao S, Pipattanasomporn M and Rahman S 2009 Challenges of PHEV penetration to the residential distribution network *2009 IEEE Power & Energy Society General Meeting* (Calgary) pp 1-8
[9] WANG HL, ZGANG YJ, MAO HP 2019 Charging Load Forecasting Method Based on Instantaneous Charging Probability for Electric Vehicles *Electric Power Automation Equipment* **3** 207-213
[10] Clement-Nyns K, Haesen E and Driesen J 2010 The Impact of Charging Plug-In Hybrid Electric Vehicles on a Residential Distribution Grid *IEEE Trans. Power Syst.* **1** 371-380.
[11] LUO ZW, HU ZC, SONG YH, YANG X, ZHAN KQ and WU JY 2011 Calculation Method of Electric Vehicle Charging Load *Automation of Electric Power Systems* **14** 36-42.
[12] LIU Q and QI ZY 2014 Load Forecasting Modeling of Electric Vehicle Based on Monte Carlo Method Electric Power Science and Engineering **10** 14-19.
[13] CHEN LD, ZHANG Y, FIGUEIREDO A 2019 Overview of Charging and Discharging Load Forcasting for Electric Vehicles *Automation of Electric Power Systems* **10** 1-15.
[14] ZHU HT, YANG X and CHEN YY 2016 Overview of the Charging Load Forecasting Methods of Plug-in Electric Vehicles *Electric Power Information and Communication Technology* **5** 44-47.
[15] Can I Fast Charge My Car *Available: https://support.fastned.nl/hc/engb/articles/204784768*

A Comparison of Learning Machines for Turboshaft Engine Gas Path Fault Pattern Recognition

Shuai Zhao[a], Jinquan Huang[b], and Feng Lu[c]

Jiangsu Province Key Laboratory of Aerospace Power System, Nanjing University of Aeronautics and Astronautics, Nanjing 210016, China

[a]1243748603@qq.com; [b]jhuang@nuaa.edu.cn; [c]lfaann@nuaa.edu.cn

Abstract. Since the operating environment of a turboshaft engine is complex and harsh, gas path components are easily damaged. It is quite important of gas path fault diagnosis for turboshaft engine from the huge accumulated data. The paper presents the data-driven ways of learning machine algorithms for gas path fault diagnosis of the engine and the performance comparisons. Various Extreme Learning Machine (ELM), including basic ELM, Kernel ELM (KELM) and Multiple Layer KELM (MLKELM), are involved. Besides, the Deep Belief Network (DBN) as one of the deep learning machines is also employed to recognize fault patterns. The simulation is carried out on a turboshaft engine with typical gas path fault modes. The results show that the MLKELM and DBN provide better classification accuracy than the ELM and KELM, and the DBN consumes less training time compared to the MLKELM.

1. Introduction

Fault diagnosis for aero-engine is the core part of engine health management system, which is also an important prerequisite to reduce maintenance cost and ensure flight safety. Aero-engine gas path fault is caused by erosion, fouling and objective damage during engine operation. As using time increases, the engine component performance degradation may also cause engine gas path components to fail [1]. Key rotating components in turboshaft engines have the higher probability to break down. Hence, it is significant to diagnose the fault of the gas path components for the turboshaft engine [2].

Up to now, researchers have proposed many methods for engine fault diagnosis. There are mainly three categories [3]: analytical model based method, knowledge based method and data-driven based method. The engine analytical model based method uses various filtering algorithms such as Kalman filter and particle filter to achieve component fault diagnosis. The data-driven based method directly relies on the engine measurement data, and no additional prior component characteristic maps and aero-thermal principle are involved [4,5]. Since learning machine algorithms owns the strong nonlinear mapping capabilities, the trained network related to learning machine algorithms might be a satisfactory candidate to represent the mathematical relationship between sensor measurement parameters and health parameters.

With the functional demands increase, more sensor measurements are utilized to gas path condition monitoring and fault diagnosis for the turboshaft engine. The available measurements enrich the fault database, and it is urgent to find the more effective way to make full use of these large-scale data to mine the valuable information for fault diagnosis.

For these purposes, several machine learning methods of ELMs and DBN are presented to the turboshaft engine. The ELM algorithm is with simple topology and fast training speed, and it is

Content from this work may be used under the terms of the Creative Commons Attribution 3.0 licence. Any further distribution of this work must maintain attribution to the author(s) and the title of the work, journal citation and DOI.
Published under licence by IOP Publishing Ltd

extremely suitable to real time classification and regression. In this paper, the basic ELM and its improved algorithm are given and applied for the engine fault pattern recognition. The MLKELM, developed from the conventional ELM, has multiple hidden layers structure with stronger feature extraction capabilities. Besides, the DBN is introduced to gas path fault diagnosis due to the more accumulated fault data. The performance comparisons of the examined algorithms are carried out using the typical fault database of the turboshaft engine.

The purpose of the paper is to find an effective way to identify gas path fault patterns. The sections are as follows: the principles of ELMs and DBN are presented for fault classification in Section 2. Section 3 gives the comparison results of the involved learning machines for gas path fault diagnosis, followed by the performance analysis with regard to classification accuracy. Finally, the conclusion is received in Section 4.

2. The Principle of learning machines and DBN

2.1 ELM, KELM and MLKELM

The Extreme Learning Machine (ELM) algorithm, developed from a single hidden layer feedforward neural network (SLFNs), was proposed by Professor Huang to overcome the shortcomings of the back propagation (BP) algorithm [6].

The ELM algorithm builds a model by randomly generating input weights $\boldsymbol{a}$, thresholds $\boldsymbol{b}$ and then output weights $\boldsymbol{\beta}$ can be obtained by solving equation $\boldsymbol{H\beta=T}$. The $\boldsymbol{T}$ is training sample output sets and feature space $\boldsymbol{H(x)}$ is defined as $\boldsymbol{H}(x)=[g(a_1^{T}x+b_1)\cdots g(a_L^{T}x+b_L)]$ where $\boldsymbol{g(\cdot)}$ is the activation function of the hidden layer. The ELM parameters do not need to be iteratively calculated and has only one hidden layer, so it has the advantages of fast training speed and simple model. However, the ELM algorithm has poor stability because of generating randomly input weights and thresholds. The kernel function instead of randomly map is used to improve the stability, where is defined as $\boldsymbol{k}(x_i, x_j)=\boldsymbol{H}(x_i)\cdot\boldsymbol{H}(x_j)$. Therefore, the KELM's output weight is received by calculating formula $\boldsymbol{\beta=H^{T}(HH^{T}+I/C)^{-1}T}$, where C, an artificially setting parameter, is usually used to balance empirical and structural risk. The MLKELM, developed from KELM, has multiple hidden layers structure and owns stronger feature extraction ability and generalization ability compared to the rest ELMs.

2.2 The DBN algorithm

Geoffrey Hinton proposed the Deep Belief Network (DBN) in 2006 [7]. Restricted Boltzmann Machine, which are basic component of DBN, has two layers neurons. The RBM network structure has been presented in figure 1. One layer neurons named visible layer ($\boldsymbol{v}$) have basically accepting a great numbers of data collected by engine sensors, the other layer neurons is mainly extracting features that was known as hidden layer ($\boldsymbol{h}$).

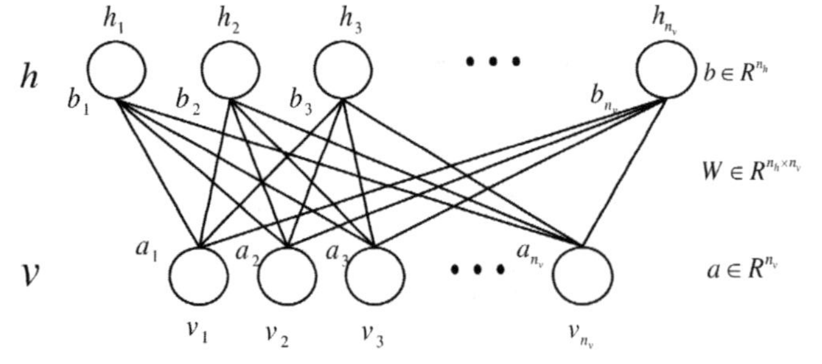

Figure 1. RBM network structure

The RBM is fully connection layers, no connection within the layer. Therefore, the neuronal states of the hidden and visible layers are independent of each other. Hence, the hidden layer h can be obtained by $P(\boldsymbol{h}|\boldsymbol{v})$ when the input is v, and then the visible layer can be obtained by $P(\boldsymbol{v}|\boldsymbol{h})$. The neuron state probability can be expressed as the following formula:

$$P(\boldsymbol{h}|\boldsymbol{v}) = \prod_{j=1}^{N} P(h_j|\boldsymbol{v})$$
$$P(\boldsymbol{v}|\boldsymbol{h}) = \prod_{i=1}^{M} P(v_i|\boldsymbol{h}) \tag{1}$$

where $\boldsymbol{v} = [v_1\ v_2\ \cdots v_M]^T$ is visible layer vector, $\boldsymbol{h} = [h_1\ h_2\ \cdots h_N]^T$ is hidden layer vector and *M, N* represent the number of the visible and hidden layer respectively.

The goal of RBM training is to find the best weight that plays a key role in the distribution. Consequently, we need to do mathematical analysis to determine the weights between the visible layer node and hidden layer node. The energy of a joint configuration can be expressed as

$$E(\boldsymbol{v},\boldsymbol{h};\theta) = -\sum_{ij} W_{ij} v_i h_j - \sum_{i} b_i v_i - \sum_{j} a_j h_j \tag{2}$$

where $\theta=\{\boldsymbol{W}, a, b\}$ are the model parameters.

Then the probability of 1 or 0 of the k-th hidden layer can be easily obtained based on the known visible layer $\boldsymbol{v}$ by factorizing equation (3). Similarly, the probability of the visible layer can also be obtained based on the known hidden layer $\boldsymbol{h}$.

$$P(h_k = 1 | \boldsymbol{v}) = sigmoid(b_k + \sum_{i=1}^{M} w_{k,i} v_i)$$
$$P(v_k = 1 | \boldsymbol{h}) = sigmoid(a_k + \sum_{i=1}^{N} w_{k,i} h_i) \tag{3}$$

DBN is composed of some restricted Boltzmann machines. The hidden layers mainly gain high-order dependence of the visible layers during the training stage. First of all, the visible vector maps to the hidden layer, and then the visible layer are reconstructed by hidden layer. The new hidden layers can obtain these new visible layers and map to the hidden layers again. After pre-training, DBNs can use labeled data and BP algorithms to fine-tune the result.

3. The principle of gas path fault diagnosis for turboshaft engine based on DBN

The figure 2 illustrates the structure of gas path fault diagnosis based on DBN. The sensor data used in this paper is generated by simulating a turboshaft engine digital model under various operating conditions.

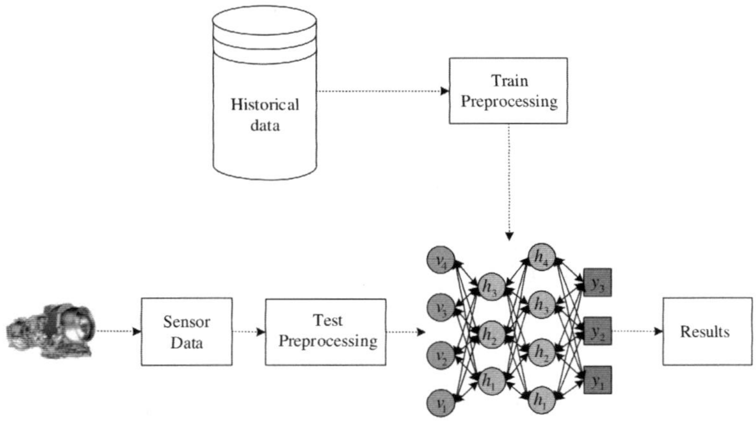

Figure 2. Structure of gas path fault diagnosis

The table 1 displayed turboshaft engine some parameters. The table is comprised of control variables, health parameters and measurement parameters. Control variables have contained fuel flow (W_f) and load torque (*angle*). Simultaneously Gaussian noise is added to the measurement parameters to simulate the actual working environment of the turboshaft engine. The maximum state correspond the relative rotational speed of the power turbine speed N_p=100%, the fuel flow W_f=0.101kg/s, load torque *angle* =10 at design point on the ground.

Particularly, health parameters *SE* and *SW* are defined as follows:

$$SE = \frac{\eta}{\eta^*}, SW = \frac{W}{W^*} \tag{4}$$

where η, η^* are actual and ideal efficiency of engine rotor component respectively; W, W^* are actual and ideal flow of engine rotor component respectively.

Table 1. Turboshaft engine parameter comparison

Control Variables	Health Parameters	Measurement Parameters
W_f- Fuel flow	SE_1- Compressor efficiency	N_p- Power turbine speed
angle- Load torque	SW_1-Compressor flow	N_g- Gas turbine speed
	SE_2- Gas turbine efficiency	T_3- Compressor outlet temperature
	SW_2- Gas turbine flow	P_3- Compressor outlet pressure
	SE_3-Power turbine efficiency	T_5- Power turbine outlet pressure
	SW_3- Power turbine flow	P_5- Power turbine outlet pressure
		T_7- Outlet temperature at the rear section of power turbine
		P_7- Outlet pressure at the rear section of power turbine

The gas path faults are divided into eleven fault modes which includes a free fault mode according to bias health parameters. The relative mode patterns of changing in health parameters is shown in table 2.

Table 2. Turboshaft engine gas path performance fault mode

Mode	Bias Health Parameter	Faulty Section
I	SE_1-1%	Compressor
II	SW_1-1%	
III	SE_1-0.7%, SW_1-1%	
IV	SE_2-1%	Gas Turbine
V	SW_2+1%	
VI	SE_2-1%,SW_2-1%	
VII	SE_3-1%	Power Turbine
VIII	SW_3-1%	
IX	SE_3-0.4%,SW_3-1%	
X	SE_3-0.6%,SW_3+1%	

The paper selected four operate points on a small rectangular envelope from engine flight envelope as a set of train data, then randomly selected a operate point in it as test data. Fault classification accuracy is calculated by ELM, KELM and MLKELM algorithms, which is expressed as the following formula:

$$Acc = \frac{N_a}{N_t} \tag{5}$$

where N_a is the number of correct classification samples, N_t is the number of overall samples.

4. Experiments and Analysis

The train data selected four operating points including *H=600m, Ma=0.1; H=600m, Ma=0.4; H=1000m, Ma=0.1; H=1000m, Ma=0.4* from the steady gas path flight envelope, then test data selected operating point *H=700m, Ma=0.2* and control variables remain unchanged. A total of four operating points were selected, then 50 new samples were obtained by adding 0.003 Gaussian noise for each fault mode.

Consequently, there are 550 testing samples and 2200 training samples.

The topological parameters involved in the three algorithms are as follows: the number of hidden layer neurons in ELM is 42; the regularization parameter of KELM is 17, the kernel parameter is 6; MLKELM has 3 hidden layers structure and the regularization parameter is [1; 50; 10], the kernel parameter is [13; 20; 1]. DBN can maintain a small network model and can have good precision when dealing with big data. Therefore, we can obtain 500 new samples by adding 0.003 Gaussian noise for each fault mode. The number of each layer of DBN, set up 2-layer structure, has 50 neurons which are used to extract sensor parameter characteristics. Its learning rate is 0.1 and the number of per batch training samples are 100.

The simulation results of the steady-state gas path fault diagnosis testing time of turboshaft engine are shown in table 3. The training time of ELMs and DBN is 0.0096s, 0.3316s, 2.6201s and 0.4813s respectively. It indicates that the DBN consumes less training time compared to the MLKELM and less testing time compared to the KELM and MLKELM.

Table 3. Gas path fault diagnosis testing time of turboshaft engine based on DBN

algorithms	I	II	III	IV	V	VI	VII	VIII	IX	X	XI	Mean
ELM	0.0022	0.0004	0.0003	0.0003	0.0004	0.0004	0.0003	0.0004	0.0004	0.0004	0.0004	0.0008
KELM	0.0054	0.0057	0.0048	0.0052	0.0057	0.0049	0.0050	0.0058	0.0050	0.0054	0.0059	0.0447
MLKELM	0.1178	0.1160	0.1157	0.1165	0.1152	0.1170	0.1155	0.1181	0.1162	0.1180	0.1148	0.2846
DBN	0.0803	0.0438	0.0020	0.0017	0.0021	0.0018	0.0019	0.0017	0.0020	0.0023	0.0021	0.0053

The simulation results of the steady-state gas path fault diagnosis classification accuracy of turboshaft engine are shown in table 4. As we can be seen from table 4, the four algorithms have a high fault diagnosis classification accuracy for the turboshaft engine envelope. Since the DBN and MLKELM has better feature extraction ability compare to rest ELMs, their mean testing accuracy which approximately reach 90% display higher. However, the ELM and KELM have poor results for the third and ninth fault mode.

Table 4. Gas path fault diagnosis classification accuracy of turboshaft engine based on DBN

Accuracy	algorithms	I	II	III	IV	V	VI	VII	VIII	IX	X	XI	Mean
Training Accuracy	ELM	0.9164	0.9100	0.8955	0.8900	0.8909	0.9191	0.8928	0.9017	0.8986	0.9047	0.8926	0.8905
	KELM	0.9427	0.8905	0.8905	0.9425	0.9486	0.9532	0.9259	0.8945	0.8955	0.8959	0.9165	0.8977
	MLKELM	0.9664	0.9732	0.9655	0.9900	0.9673	0.9982	0.9659	0.9659	0.9627	0.9959	0.9686	0.9700
	DBN	0.9705	0.7680	0.9375	0.9995	0.9985	0.9965	0.9845	0.8380	0.8950	0.8980	0.8465	0.8997
Testing Accuracy	ELM	0.8800	0.8800	0.7200	0.9520	0.9160	0.9600	0.9920	0.8960	0.8000	0.9800	0.8000	0.8109
	KELM	0.9600	0.9000	0.5800	0.9200	0.9000	0.9400	0.9000	0.9800	0.5400	0.9800	0.8800	0.8091
	MLKELM	0.9600	0.8400	0.9400	0.9800	0.9600	0.8800	0.9000	0.9200	0.8200	0.9800	0.9600	0.9164
	DBN	0.9620	0.8360	0.8320	0.9980	0.9960	0.9920	0.9500	0.9820	0.9820	0.9300	0.8500	0.9102

5. Conclusion

The paper introduced a method that the deep belief network is applied to the gas path fault patterns for turboshaft engine. The objective is to obtain engine steady-state fault diagnosis results through sensor data that indicates the working condition of the turboshaft engine. The engine's health parameters are offset to obtain the corresponding fault mode, then the 8 sensor parameters was selected to train the DBN network model to recognize the fault identification accuracy in the selected operating points of engine flight envelope. The simulation results show that the DBN is superior to the rest ELMs for engine gas path performance fault identification with regard to stability and correct rate.

References

[1] Mercer C R, Simon D L and Hunter G W 2007 NASA-TM-2007-0022364
[2] Kong X X and Wang X 2012 *J. Aerosp. Power.* **27** 2599-2608

[3] Thomas P, Fast M and Thern M 2011 *J. Appl. Energ.* **88** 3898-3904
[4] Yan X , Jia M and Xiang L 2016 *J. Meas. Sci. Technol.* **27** 075002
[5] Yang K, Shan G and Zhao L 2006 Proc. Conf. Intelligent Control and Automation vol 1
[6] Huang G B and Chen L 2008 *J. Neurocomputing.* **71** 3460-68
[7] Hinton G E, Osindero S and The Y 2006 Neural Computation

IOP Conf. Series: Materials Science and Engineering **752** (2020) 012014 doi:10.1088/1757-899X/752/1/012014

Optimal Configuration of Fault Current Limiter Based on Multi-objective Decision

Wenxiong Mo[1,a], Yong Wang[1,b], Jun Chen[1,c], Shengya Qiao[1,d], Lu Zhu[1,e] and Rong Fu[2,f]

[1] Guangzhou Power Supply Bureau Co., Ltd, Guangzhou 510620, Guangdong Province, China
[2] School of Electrical Engineering and Automation, Wuhan University, Wuhan 430072, Hubei Province, China

[a]gzmwx@139.com; [b]wangy@guangzhou.csg.cn; [c]laochen.locky@163.com; [d]396349222@qq.com; [e]315995738@qq.com; [f]furong96@outlook.com

Abstract. To perfect the application of the fault current limiter in grid and improve the use efficiency, an optimal configuration method of fault current limiter based on multi-objective decision is proposed. Firstly, use rate of current change to select alternative branches, then taking the cost and current-limiting effect of fault current limiter as the targets, considering installation location, quantity of fault current limiter comprehensively, using α-method to determine the weight coefficient, establish multi-objective optimization model, and use immune algorithm to obtain the optimal scheme of branch installation. Next, by comparing the configuration of fault current limiter in the bus connection, the best configuration scheme is obtained. Finally, this method is verified by analyzing the IEEE39 node standard network.

1. Introduction

As the load level continues to rise, the scale of power system continues to increase, the short-circuit current has become one of the main factors that endanger the stability, safety and reliability of power systems [1-2]. Therefore, Fault Current Limiter, which can limit the short-circuit current without affecting the normal operation of the power system, has important research significance.

The key to optimize the fault current limiter is to configure the number, impedance and installation position of the FCL with the goals of reliability and economy [1].

The references[2-5] optimizes the type, the use scale , the installation range and the impedance value of the fault current limiter; In reference [6], a multi-objective optimization model is established based on system power flow constraints, and the incomplete enumeration method and discrete particle swarm optimization algorithm are used to optimize the configuration of short-circuit current limiting for large-scale and complex networks; In reference [7], based on the immune algorithm, Pareto multi-objective optimization is used to study the configuration of high temperature superconducting fault current limiters; In reference [8], the fast-non-dominated sorting genetic algorithm is used to obtain a multi-objective optimization model based on multi-objective decision and Pareto optimal solution. All of the studies above have been optimized the installation of the fault current limiter in the branch, but the bus connection has not been further considered.

Based on the above research, this paper proposes an optimal configuration method of fault current limiter based on multi-objective decision. The short-circuit current rate-of-change sensitivity analysis is

Content from this work may be used under the terms of the Creative Commons Attribution 3.0 licence. Any further distribution of this work must maintain attribution to the author(s) and the title of the work, journal citation and DOI.
Published under licence by IOP Publishing Ltd

used for branch screening. The installation position, installation quantity and current limiting effect of the fault current limiter are comprehensively considered. The fault current limiter impedance value and short-circuit current of node are constrained. With the goal of cost and current limiting effect, the weighting coefficient is determined by the α-method, and the multi-objective optimization model is established. Then, the optimization result of the branch installation is obtained by using the immune algorithm, and the classification enumeration method is used to optimize the configuration of the fault current limiter in the bus connection. Finally, analyze and compare both schemes to get the best one.

2. Mathematical model and installation location selection of FCL

2.1. Mathematical model of FCL

The High Coupled Split Reactor-Fault Current Limiter (HCSR-FCL) used in the paper working principle is shown in figure 1.

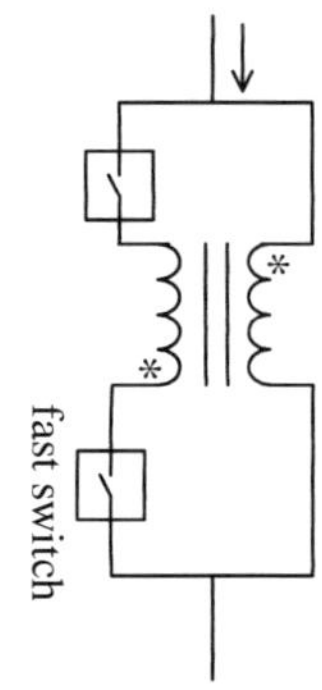

Figure 1. The principle diagram of HCSR-FCL

The HCSR-FCL consists of two coils that are coupled to each other. The two coils are short-circuited at one end, two quick switches on the same side is designed to eliminate the potential difference between the two splitting coils. When the two branches current is balanced, the magnetic flux generated by the two coils cancel each other out in the common magnetic circuit, and only shows a small leakage inductance; when the two branches are not balanced or only a single branch is on, the magnetic flux generated by the coil is large, which can limit the fault current.

According to the principle of short-circuit calculation, when the three-phase short-circuit fault occurs at node *k*, its short-circuit current is:

$$I_k^* = \frac{1}{Z_{kk}^* + z_f^*} \tag{1}$$

Where , Z_{kk}^* is the self-impedance of node *k*; z_f^* is the transition impedance of node *k*; I_k^* is the short- circuit current of node *k*.

When HCSR-FCL is put into operation, use additional branch method to modify the node impedance matrix. As shown in figure 2, series impedance Z_{FCL} can be equivalent to impedance Z_F in parallel, which can be calculated by equation (2).

$$Z_F = (-Z_{pq}) // (Z_{pq} + Z_{FCL}) = (-Z_{pq}) \frac{(Z_{pq} + Z_{FCL})}{Z_{FCL}} \tag{2}$$

Then, after the short-circuit fault occurs, when the single-arm impedance of HCSR-FCL is Z_{FCL} , the self-impedance of any node k is expressed as:

$$Z_{kk}^{'} = Z_{kk} - \frac{(Z_{kp} - Z_{kq})^2}{Z_{pp} + Z_{qq} - 2Z_{pq} + Z_F} = Z_{kk} + \Delta Z_{kk} \tag{3}$$

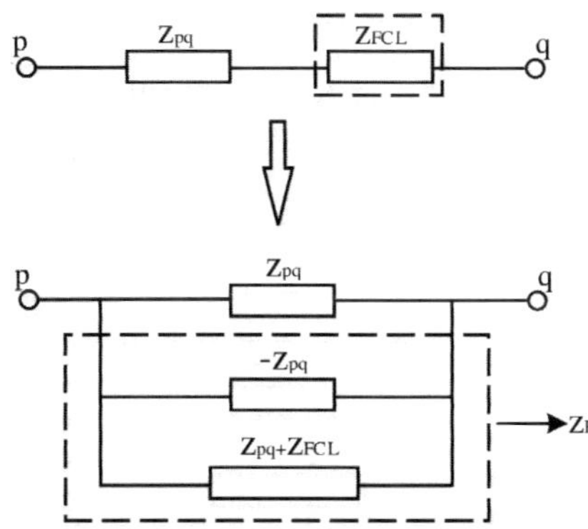

Figure 2. The equivalent impedance of HCSR-FCL

Where Z_{kk} is the self-impedance of node k before HCSR-FCL put into operation, ΔZ_{kk} is the change of node *k*'s self-impedance after HCSR-FCL putting into operation.

2.2. Screening based on sensitivity analysis
In the large transmission and distribution network, in order to reduce the amount of calculation and avoid falling into the "dimensional disaster", a sensitivity analysis method can be used to screen out a certain range in advance.

References [4,7] proposed two methods of sensitivity analysis, namely the sensitivity of the node impedance increment based on the branch impedance increment and the sensitivity of short-circuit current rate-of-change. In the actual comparative study, the sensitivity of short-circuit current rate-of-change has shorter computation time and more operability when calculating the current limiting effect on multiple nodes. Therefore, in this paper, this method is used to screen the branch.

When the three-phase short-circuit fault occurs in the grid, the short current of node *k* is I_k, and the short-circuit current rate-of-change node *k* after the FCL is put into the branch *p-q* is

$$\begin{aligned} C_{pq}^{k} &= \frac{\Delta I_k}{I_k} = \left(\frac{U_k^{(0)}}{Z_{kk} + \Delta Z_{kk}} - \frac{U_k^{(0)}}{Z_{kk}} \right) / \left(\frac{U_k^{(0)}}{Z_{kk}} \right) \\ &= -\frac{\Delta Z_{kk}}{Z_{kk} + \Delta Z_{kk}} \end{aligned} \tag{4}$$

In equation (4), C_{pq}^{k} is the short current rate-of-change after the fault current limiter is installed on the branch p-q; ΔI_k is the short current change amount of the node *k*; Z_{kk} is the self-impedance of the node *k* before the FCL is installed; ΔZ_{kk} is the self-impedance change amount of the node *k* after the FCL is installed.

The equation (4) can be used to calculate the short current rate-of-change of each branch to the node. According to the value, the first few branches can be selected as the candidate branch, and the optimal configuration of FCL can be performed on this basis.

3. Optimization configuration of FCL

3.1. Objective function and constraint conditions
From the overall effect of the network, the cost calculation of HCSR-FCL is mainly affected by the installed quantity and the total impedance of the FCL. Therefore, the cost function of the HCSR-FCL is as follows:

$$f = 2\square\sum_{i=1}^{N} Z_{FCL} + \omega N_{FCL} + F_a \tag{5}$$

In equation (5), Z_{FCL} is the impedance value of a coil in the HCSR-FCL, one HCSR-FCL contains two identical coils; ω is the installation cost coefficient of the HCSR-FCL; N_{FCL} is the total number of installations of the HCSR-FCL on the branch; F_a is the penalty function setting to ensure the short current of each node can be limited to the allowed value after the HCSR-FCL is put into operation.

In order to facilitate the calculation of the objective function, a current-limiting cost sub-function is proposed:

$$F_1 = \frac{1}{f} = \frac{1}{2\square\sum_{i=1}^{N_{FCL}} Z_{FCL}(i) + \omega N_{FCL} + F_a} \tag{6}$$

The current limiting effect of each node can be determined according to the ratio of the difference between the short-circuit current value and the target current value. In equations below, η_k is the current limiting effect on the node k after the fault current limiter is put into operation; I_G is the target current; $I_k^{'}$ is the short current of the node k after the fault current limiter is put into operation; Z_{Gk} is the self-impedance of the node k under the target current; $U_k^{(0)} = 1$ when calculating in standard value.

$$\begin{aligned}\eta_k &= \frac{I_G - I_k^{'}}{I_G} = \left(\frac{U_k^{(0)}}{Z_{Gk}} - \frac{U_k^{(0)}}{Z_{kk}^{'}}\right) / \left(\frac{U_k^{(0)}}{Z_{Gk}}\right) \\ &= \frac{Z_{kk}^{'} - Z_{Gk}}{Z_{kk}^{'}}\end{aligned} \tag{7}$$

$$I_k^{'} = \frac{U_k^{(0)}}{Z_{kk}^{'}} \tag{8}$$

$$Z_{Gk} = \frac{U_k^{(0)}}{I_G} \tag{9}$$

When there is only one excessive node in the system, equation (7) is the current-limit effect sub-function; when there are multiple excessive nodes, the current-limit effect sub-function of the FCL is

$$F_2 = \sum_{k=1}^{\Gamma} \varepsilon_k \eta_k \tag{10}$$

Where, Γ is a set of all nodes in the system that do not meet the requirements; ε_k is a weight coefficient, which is used to balance the current limiting effect between excessive node and insufficient margin node . When the original short-circuit current of the node k is greater than I_G, the node k is an excessive node, and $\varepsilon_k = 1$; when the original short current of the node k is smaller than I_G but greater than margin of short-circuit current I_M, the node k is an insufficient margin node, and $\varepsilon_k = I_k / I_G$.

In the multi-objective mathematical model, in addition to the objective functions and decision variables, the optimization of the fault current limiter configuration must satisfy several constraints:

(1) Fault current limiter impedance value constraint

The impedance value that HCSR-FCL can be put into the line when a short circuit fault occurs must meet:

$$0 \le Z_{FCL}(i) \le Z_{\max} \tag{11}$$

In equation (11), $Z_{FCL}(i)$ is the fault current limiter impedance value that can be input on each candidate branch; $Z_{\max}$ is the maximum impedance value of the fault current limiter that is allowed to be input on each branch and to prevented the cost from being excessive.

(2) Short-circuit current constraints of nodes in the power grid

The purpose of installing the FCL is to limit the short-circuit current of the node of the power grid to a suitable range. The constraints are as follows:

$$I_k^{'} < I_G \tag{12}$$

In addition, there should be no isolated nodes in the power grid, and the active power and reactive power in the power flow calculation must be balanced.

Therefore, the mathematical model of FCL configuration optimization based on multi-objective decision is as follows:

$$F = \begin{cases} \max(F_1) \\ \max(F_2) \\ 0 \le Z_{FCL}(i) \le Z_{\max} \\ I_k^{'} < I_G \end{cases} \tag{13}$$

3.2. α-method to determine the weight coefficient of the multi-objective optimization mathematical model

According to reference [9], the weight coefficient can directly reflect the degree of importance of the objective function. It can be known from the definition of the fault current limiter cost sub-function and the current limiting effect sub-function that the larger the value of the current limiting cost sub-function, the smaller the value of the current limiting effect sub-function. Obviously, there is no absolute optimal solution for the multi-objective optimization mathematical model. Therefore, this paper chooses to use the α-method to determine the weight coefficient of the multi-objective optimization mathematical model.

Firstly, introduce an overall objective function F:

$$F = \lambda_1 \times f_1 + \lambda_2 \times f_2 \tag{14}$$

In equation (14), F is the function value for determining the energy saving potential of the line; λ_1、λ_2 is the weight coefficient, $\lambda_1 + \lambda_2 = 1, \lambda_1 > 0, \lambda_2 > 0$; f_1 is the cost sub-function value of the FCL, and f_2 is the current limiting effect sub-function value of the FCL.

Set the optimal solution for the single-objective problem $(P_l)\max_{x\in R} f_i(x)$ $(i=1,2;x\in R)$ as x^i $(i=1,2)$, remark:

$$\begin{cases} f_1(x^1) = f_1^1 \\ f_2(x^1) = f_2^1 \\ f_1(x^2) = f_1^2 \\ f_2(x^2) = f_2^2 \end{cases} \tag{15}$$

As shown in figure 3, in the image space, make a straight line over the point (f_1^1, f_2^1) and (f_1^2, f_2^2).

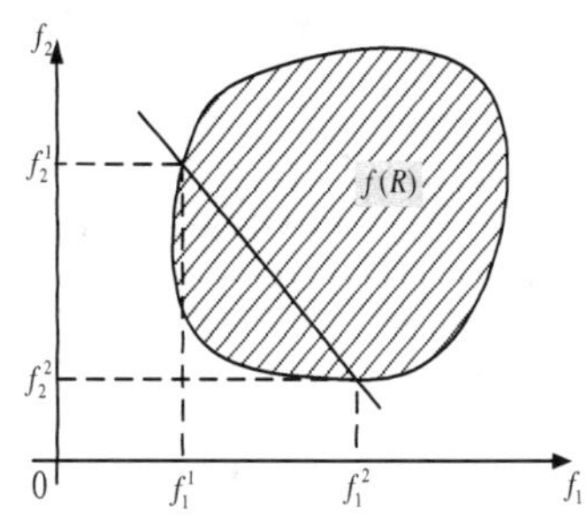

Figure 3. The Schematic diagram of the method for determining the weight coefficient

Let the linear equation be

$$\lambda_1 f_1 + \lambda_2 f_2 = \alpha \tag{16}$$

Then there is

$$\begin{cases} \lambda_1 + \lambda_2 = 1 \\ \lambda_1 f_1^1 + \lambda_2 f_2^1 = \alpha \\ \lambda_1 f_1^2 + \lambda_2 f_2^2 = \alpha \end{cases} \tag{17}$$

The premise of the α-method is that there is no absolute optimal solution, that is, the optimal optimization result cannot be directly obtained by comparing the two sub-objective functions. Therefore, there must be

$$\begin{cases} f_1^2 > f_1^1 \\ f_2^1 > f_2^2 \end{cases} \tag{18}$$

Solving equation (17) under the condition of equation (18), we can get equation (19) for solving two weight coefficients:

$$\begin{cases} \lambda_1 = (f_2^1 - f_2^2)/[(f_1^2 - f_1^1) + (f_2^1 - f_2^2)] \\ \lambda_2 = (f_1^2 - f_1^1)/[(f_1^2 - f_1^1) + (f_2^1 - f_2^2)] \\ \lambda_1 > 0, \lambda_2 > 0, \lambda_1 + \lambda_2 = 1 \end{cases} \tag{19}$$

3.3. Optimization steps based on immune algorithm

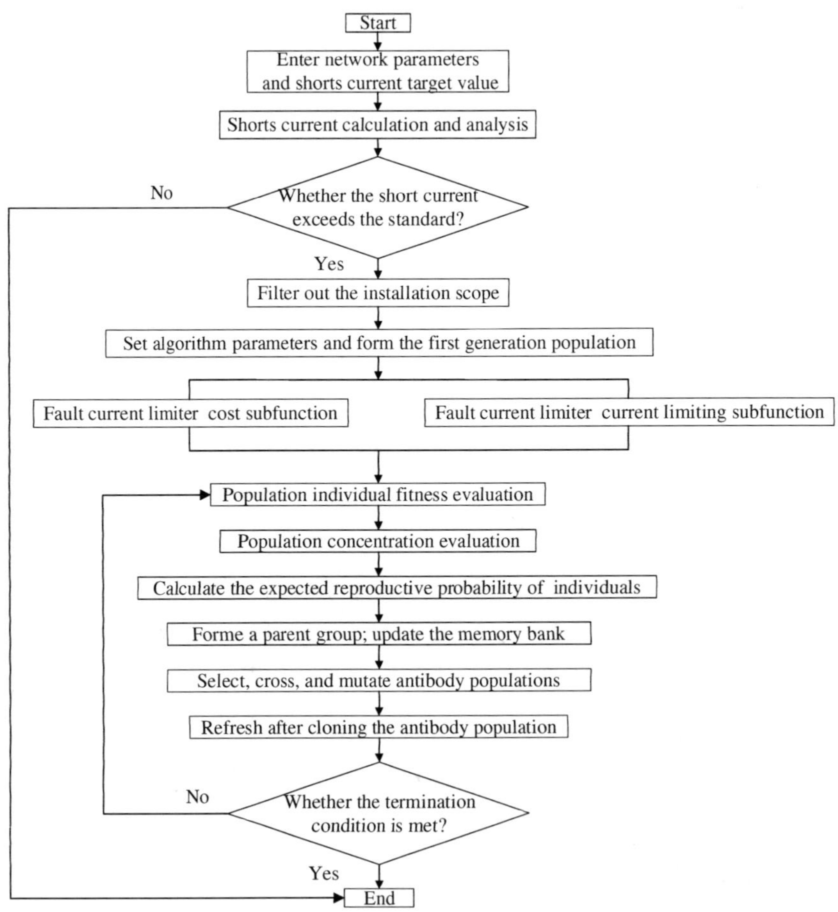

Figure 4. FCL optimization configuration flow chart based on AIA

In this paper, the immune algorithm is used to optimize the multi-objective configuration of the FCL. The scheme realizes the comprehensive optimization of the two sub-objective functions.

The specific operation steps of the optimization process based on the immune algorithm is shown in figure 4.

The gene length of the antibody in the immune algorithm is the number of candidate branches. The gene coding of the antibody indicates the installation number of the FCL, and the position of the gene is the installation position of the FCL. Others are immune algorithms [10-11] normal operation, so it won't be gone into details.

4. Case analysis

4.1. Simulation case

This paper takes IEEE39 nodes system as an example, as shown in figure 5, for FCL optimization configuration. Let the target value of the short-circuit current I_G=84pu. The margin of the short-circuit current is 10% of the target value, and the margin value of the short current I_M=75.6 pu.

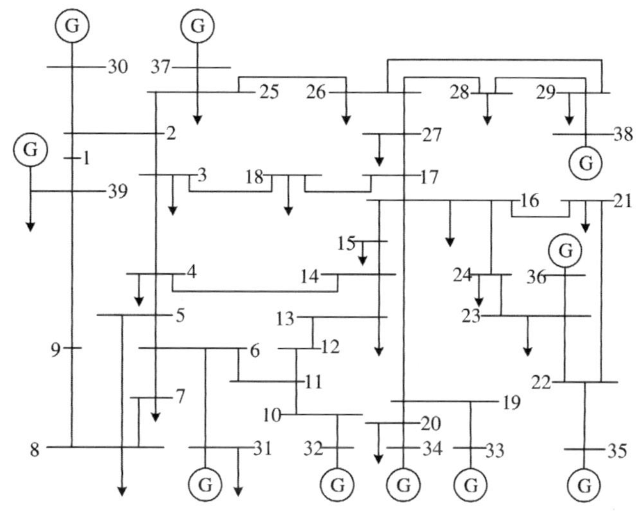

Figure 5. IEEE39 nodes system

When the three-phase short circuit fault occurs, the short-circuit current exceeds the standard and the nodes that do not meet the margin requirements are shown in Table 1.

It is apparent from the table that the three-phase short-circuit current of the node 39 is particularly large, because the transient reactance of the generator connected to the node 39 is extremely small, so using the HCSR-FCL to limit its short current is not economical in actual operation, and the effect is not good. It is more suitable to limit the short current of node 39 by means of disconnection [12]. Therefore, this paper only studies the FCL optimization for the Node 2, Node 16, and Node 30.

Table 1. Excessive nodes and insufficient margin nodes

Node number	I_k /pu	Excessive / Insufficient margin
2	85.2	excessive
16	82.4	insufficient margin
30	79.1	insufficient margin
39	280.9	excessive

IOP Conf. Series: Materials Science and Engineering **752** (2020) 012014 doi:10.1088/1757-899X/752/1/012014

4.2. Branch screening

According to the HCSR-FCL start condition, the sensitivity of the short-circuit current rate-of-change on each node is obtained when the FCL installed on each branch is operated.

The magnitude of the short-circuit current rate-of-change of the node that does not meet the requirement is sorted, and each node is analysed based on the first 6 bits, as shown in Table 2. Exclude duplicate branches to get the final installation range, as shown in Table 3.

Table 2. Branch screening based on sensitivity of short current change rate

Node that does not meet the requirements	Branches (rate -of-change)
2	1-2(1.76%);1-39(1.76%);2-3(4.22%); 2-25(3.94%);2-30(5.55);25-37(1.56%)
16	14-15(2.12%);15-16(2.12%);16-17(4.44%); 16-19(3.52%);16-21(1.71%);19-33(1.97%)
30	1-2(0.44%);1-39(0.44%);2-3(1.07%); 2-25(1.00%);2-30(6.81%);25-37(0.39%)

Table 3. Range based on short-circuit current rate-of-change sensitivity

Branches name
1-2; 1-39; 2-3; 2-25; 2-30; 14-15; 15-16; 16-17; 16-19; 16-21; 19-33; 25-37

4.3. HCSR-FCL configuration optimization on branch

According to the content above, the FCL configuration is optimized based on the immune algorithm. The parameters of the immune algorithm are set as follows: the population size is set to 400, the memory capacity is set to 100, the maximum number of iterations is set to 5000, the gene length of the antibody is set to 12 ,the penalty function is set to 10000, the crossover probability is set to 0.5, the mutation probability is set to 0.4, and the diversity evaluation parameter is set to 0.95.

For economical and practical considerations, we only install one HCSR-FCL on each branch. Because the impedance of HCSR-FCL is fixed during production, and consider the current limiting range and give the impedance value constraint.

$$Z_{FCL}(i)_{\max} = 0.0084\,pu \tag{20}$$

According to reference [8], take ω=5. Taking the cost sub-function F_1 as the horizontal axis and the current limiting effect sub-function F_2 as the vertical axis, the partial results obtained are shown in figure 6.

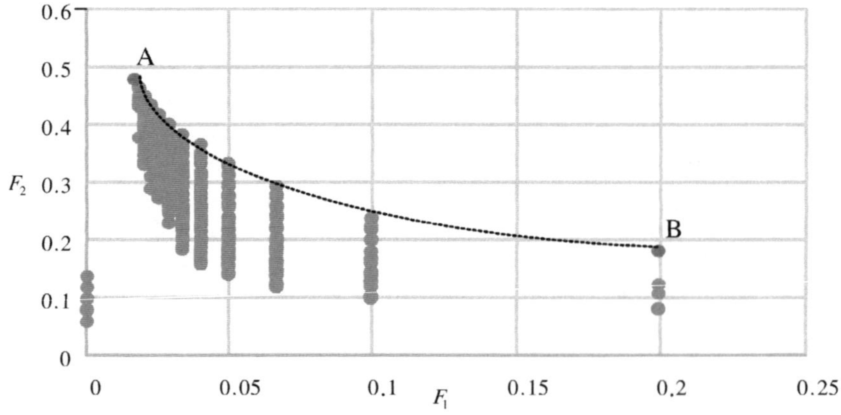

Figure 6. HCSR-FCL configuration optimization results in the grid (n = 0,1)

It can be seen from the figure that the cost sub-function F_1 of HCSR-FCL and the HCSR-FCL current limiting effect sub-function F_2 are inversely proportional to the general trend. In order to obtain the appropriate weight, according to the equation (19)

$$\begin{cases} \lambda_1 = (f_2^1 - f_2^2)/[(f_1^2 - f_1^1) + (f_2^1 - f_2^2)] = 0.685 \\ \lambda_2 = (f_1^2 - f_1^1)/[(f_1^2 - f_1^1) + (f_2^1 - f_2^2)] = 0.315 \end{cases} \tag{21}$$

Substituting λ_1 and λ_2 into equation (14) to perform optimization calculation of multi-objective model. Finally, the fault current limiter configuration scheme obtained is shown in Table 4. In this case, the cost sub-function $F_1 = 0.1993$, the current limiting effect sub-function $F_2 = 0.1803$. after the FCL is put into operation, the total value of the drop of short-circuit current is 10.54 kA, and the short-circuit current of the node is shown in figure 7.

Table 4. Optimization configuration of FCL

Branch	Number of FCLs	Branch	Number of FCLs
1-2	0	15-16	0
1-39	0	16-17	0
2-3	0	16-19	0
2-25	0	16-21	0
2-30	1	19-33	0
14-15	0	25-37	0

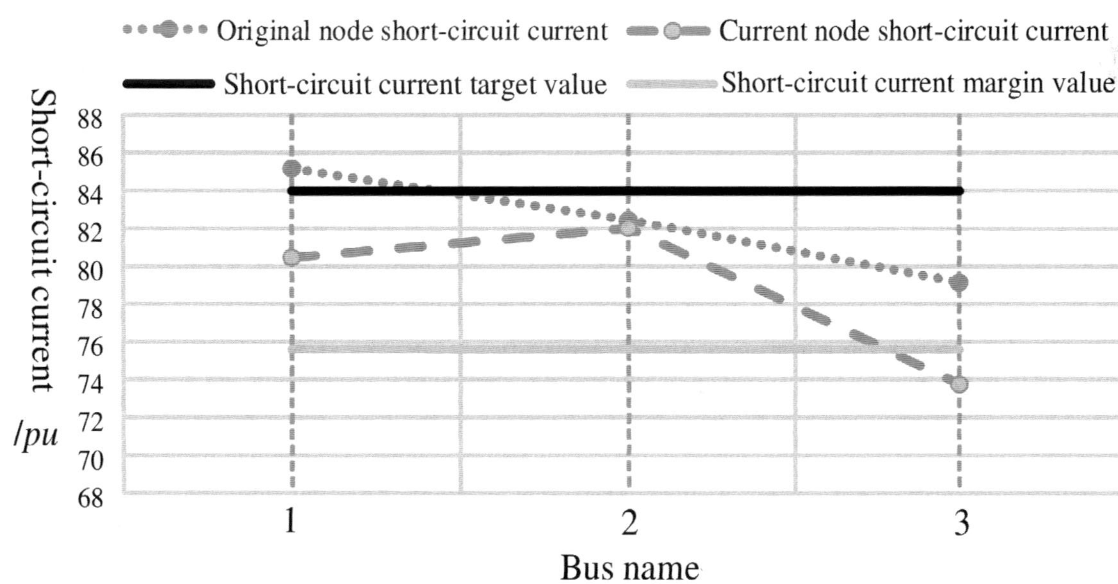

Figure 7. Optimization scheme (Z_{FCL}=0.0084pu)

4.4. HCSR-FCL configuration optimization on bus

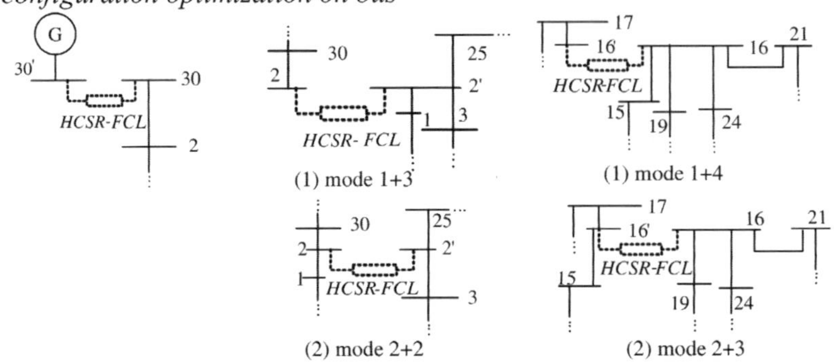

(a) HCSR-FCL on bus 30 (b) HCSR-FCL on bus 2 (c) HCSR-FCL on bus 16

Figure 8. Fault current limiter installed on bus 2, bus 16 and bus 30

It is assumed that the following bus lines have been split, that is, the cost of the splitting bus lines is not considered, and the cost function is only affected by the impedance value of the fault limiter and the number of installations. To contrast with the branch configuration schemes, consider that one fault current limiter is installed on bus 2, bus 16 and bus 30 respectively (in this case, the cost can be considered the same due to the fixed impedance and quantity of the FCL), as shown in figure 8.

Table 5. The connection between the original node/the newly added node and other nodes

(a)Bus 30

Scheme no.	30'	30
1	G	2

(b)Bus 2

Scheme no.	2'	2
2	25,30	1,3
3	3,30	1,25
4	1,30	3,25

(c)Bus 16

Scheme no.	16'	16	Scheme no.	16'	16
5	15,17	19,21,24	10	17,21	15,19,24
6	15,19	17,21,24	11	17,24	15,19,21
7	15,21	17,19,24	12	19,21	15,17,24
8	15,24	17,21,24	13	19,24	15,17,21
9	17,19	15,21,24	14	21,24	15,17,19

Table 6. Effect of different installation schemes

(a)Install a FCL on Bus 30

Bus name	2	16	30	30'
before	85.18376	82.43575	79.14378	--
after	85.17086	82.41231	79.14111	47.53833

(b)Install a FCL on Bus 2

Bus name		2	2'	16	30
before		85.18376	--	82.43575	79.14378
	2	71.53226	76.58278	81.97229	77.02138
after	3	71.1357	76.93333	82.28842	77.11471
	4	**72.92076**	**74.98176**	**81.46842**	**76.00746**

(c)Install a FCL on Bus 16

Bus name		2	16	16'	30
before		85.18376	82.43575	--	79.14378
after	5	83.93486	72.94782	70.6405	78.85613
	6	84.9626	74.39849	69.48991	79.09353
	7	85.08196	76.92309	67.13695	79.12082
	8	85.13495	78.33022	64.89294	79.13292
	9	84.87238	71.16274	72.91254	79.07286
	10	84.70694	74.14276	70.72133	79.03488
	11	84.54515	75.82152	68.66815	78.99763
	12	84.44876	75.14124	69.06493	78.97538
	13	84.61905	76.74855	66.89652	79.01466
	14	84.79545	63.01515	78.48036	79.05522

The preliminary analysis of any of the above two modes on the bus 2 and 16 is carried out. The results show that the bus 2 has a better current limiting effect in mode 2+2, and the bus 16 has a better current limiting effect in mode 2+3. Finally, the classification enumeration method is used. The specific scheme is shown in Table 5. The current limiting effect of installing an FCL on the bus connection is shown in Table 6.

From the table data, the method of installing HCSR-FCL on bus 2 using scheme 4 enables the short-circuit current of all nodes (except node 39, not considered) to be limited to the short-circuit current target value, and the total number of node short-circuit current drops is the most. It is the most economical and effective HCSR-FCL configuration measure among the above schemes. Compared with the optimal branch configuration scheme in Section 4.3, it can be found that both schemes use only one HCSR-FCL for node short-circuit current limitation. However, the scheme of installing the HCSR-FCL on the bus 2 connection has more total short-circuit current drops and better current limiting effect. Therefore, the scheme shown in Figure 8(b)(2) which install the HCSR-FCL on the bus 2 is the optimal configuration scheme in IEEE39 nodes standard system.

5. Conclusion

This paper establishes a mathematical model for HCSR-FCL configuration optimization in the power grid. Considering the HCSR-FCL impedance and the node short-circuit current constraint installed on the branch, the artificial immune algorithm with faster search speed and stronger global search ability than genetic algorithm is used to determine the weight coefficient by α method. The cost sub-function and the current limiting effect sub-function are used to perform multi-objective optimization analysis on the configuration of HCSR-FCL on the grid branch. On this basis, using the classification enumeration method, analyse the different installation position of HCSR-FCL on the bus connection, and the optimal configuration scheme of the HCSR-FCL on the bus is obtained.

The simulation verification of the IEEE39 nodes system shows that the optimal configuration scheme on the branch is more convenient to use. However, for complex networks with many excessive and insufficient margin nodes, it may be necessary to install HCSR-FCL on more branches. Compared with the bus connection installation scheme, it is not economic relatively. For the installation of HCSR-FCL on the bus connection, the short-circuit current of multiple branches can be limited at the same time. The number of FCLs used is less, and the current limiting cost is lower, but the optimal scheme needs to be obtained by the method of classification enumeration, which is complicated for the optimization configuration of complex grid.

Acknowledgments

This work was supported by National Key R&D Project Program "Smart Grid Technology and Equipment" key project "Development of Economical High-voltage AC Current Limiter of 500kV and Above Voltage Grade"(Item no. 2018YFB0904300)

References

[1] Linzhi Y, Tianqi L and Jianquan W 2018 Global Optimization Configuration Algorithm for Ultra High Voltage Fault Current Limiter Based on Modified Particle Swarm Optimization *Electric Power Automation Equipment* **38(02)** 145–152

[2] Jianchao Z 2014 Development Status and Application Prospect of Fault Current Limiter *Proceedings of the CSEE* **34(29)** 5140–48

[3] Chunming T, Shu D, Cheng G, Chengzhi W, Peng Y and Fei J 2014 Analysis of New Multifunctional Solid State Current Limiter and Its Filter Selection *Power System Technology* **38(06)** 1634–38

[4] Wenwang H, Zhinong W, Guoqiang S, Qun L and Jiankun L 2012 Optimal Configuration of Superconducting Fault Current Limiter Based on Sensitivity Method *Automation of Electric Power Systems* **36(22)** 62–67

[5] Ping J, Jiayi M and Wei H 2011 Optimal Configuration of Series Reactors Considering Transient Stability Constraints *Electric Power Automation Equipment* **31(11)** 38–42+55

[6] Lili C, Minxiang H, Nuo X, Deqiang G and Si Z 2010 Optimal configuration of current limiting measures considering power flow constraints *High Voltage Engineering* **36(06)** 1572–76

[7] Baichao C, Liangliang W, Yang L, Zhikang T, Jiayu Y, Feng C, Dexin N and Jing P 2015 Pareto multi-objective optimization configuration of high temperature superconducting fault current limiter based on immune algorithm *Power System Technology* **39(05)** 1343–50

[8] Qinyong Z, Dong Y, Yutian L and Yong T 2015 Multi-objective optimization of multi-DC feed-in grid limiting short-circuit current scheme *Automation of Electric Power Systems* **39(03)** 140–145+151.

[9] Yiyun L and Jiali D 1992 *Method and theory of multi-objective optimization* Jilin Education Press.

[10] Xinzi J, Kezong T and Shang G 2008 Hybrid algorithm of ant colony algorithm and immune algorithm[J]. *Science Technology and Engineering* **05** 1327– 30.

[11] Rirong Z and Zongyuan M 2003 An Improved Artificial Immune Algorithm[J]. *Computer Engineering and Applications* **33** 55–57.

[12] Jiaxin Y, Jianbing P and Wenli F 2013 An immune- algorithm-based space-vector PWM control strategy in a three-phase inverter *IEEE Trans on Industrial Electronics* **60(5)** 2084– 93.

[13] Qingren J, Hongkun C, Zhiping Y 2011 Overview of Installation Location of Fault Current Limiter in Power System *Electric Power Construction* **32(02)** 52–55.

State Evaluation for Intelligent Distribution Terminal Units Based on Mining Association Rules

L Zhu[1], Y Ge[1], X D Song[2,*], W B Zhao[3] and J C Guo[4]

[1]Guangdong Power Grid Co. Ltd., Guangzhou, Guangdong 510000, China
[2]Guangdong Diankeyuan Energy Technology Co. Ltd., Guangzhou, Guangdong 510080, China
[3]Zhongshan Power Supply Bureau of Guangdong Power Grid Co. Ltd., Zhongshan, Guangdong 528400, China
[4]School of Electrical Engineering, Southeast University, Nanjing, Jiangsu 210096, China

*694011613@qq.com

Abstract. The construction of Ubiquitous Power Internet of Things expands the content and scale of data in distribution network. In order to meet the demand of intelligent operation and maintenance of distribution terminals, this paper proposes a method of mining association rules in data to realize evaluation of distribution terminal units. In this method, firstly, the Apriori algorithm with directional constraint of terminals' functional module is used to mine the correlation indicators system under each divided functional module of terminals. Secondly, the comprehensive measurement of confidence and lift degree is used to increase the objectivity of evaluation's weights of indicators which found by mining associations rules. Then, combined with the fuzzy comprehensive evaluation method, according to the correlation indicators system, the status evaluation of intelligent distribution terminals is realized. Finally, the historical status data of the actual disabled terminals are selected to verify the good effectiveness of the method that described in this paper.

1. Introduction

Distribution terminal units [1], as the important monitoring and control equipment in the process of intelligent power grid, have a direct impact on the reliability and security of the distribution network. There are many types of distribution terminal units, which are widely distributed in complex operating environment and dropped into abnormal state without perception in time [2]. The traditional planned maintenance methods, which are involved in study [3], are inefficient and difficult to meet the operation and maintenance requirements of distribution terminal units. Therefore, it is necessary to study effective methods of online status assessment and diagnosis of terminals, and to find out the abnormal running state timely and accurately, so as to avoid the deterioration of the state of terminals.

At the present stage, benefit by the development of the Ubiquitous Power Internet of Things [4], data collection, data mining and other technologies [3], in the distribution information system, the scale of data related to the operation of distribution equipment in grid continues to expand. In the context of big data, how to effectively collect, fuse, mine and utilize multi-source and heterogeneous data to realize the evaluation of power equipment and system state is a hot research direction at present.

Content from this work may be used under the terms of the Creative Commons Attribution 3.0 licence. Any further distribution of this work must maintain attribution to the author(s) and the title of the work, journal citation and DOI.
Published under licence by IOP Publishing Ltd

Literature [5] evaluated the state of distribution terminals based on D-S evidence theory, but only selected simple statistics such as the off-line's frequency of terminal equipment as the state characteristic. The selection of indicators was too conservative, which may affect the credibility of results of state evaluation. Literature [6] divides the multi-dimensional factors that affect the operation state of distribution network into main factors and secondary factors, and constructs a recursive model to evaluate the operation state of distribution network combined with analytic hierarchy process. However, the determination of indicators' weights is solely dependents on a small number of experts' judgement and mark, so how to determine the objective weights of data analysis is its improvement direction. Literature [5-6], all based on the comprehensive fuzzy evaluation method, respectively carried out improvement research on the state evaluation methods of mine hoist and relay protection. They all have good results in a certain test range, but the establishment process of state indicators lacks the process of data mining and analysis, which means the indicators is not established objective enough. Literature [7-8] quantifies the correlation between transmission lines' characteristic factors and lines' lightning strike faults with association rules, and establishes a lightning damage risk assessment model based on evidence theory, which is of reference significance for mining state data of distribution terminals' association rules.

This paper proposes a terminal state evaluation method based on data mining, aiming at the problems of complicated distribution terminal modes, hardware standardization objects that are difficult to establish state evaluation systems, difficult to determine state indicators under complex data, and the shortcomings of traditional evaluation methods, such as fixed mode and insufficient objectivity. First, the terminal equipment is divided into modules according to standardized functions, so as to establish the evaluation object of the device layer and the functional layer. Then, based on the Apriori algorithm with the constraints adding a single front and back item, the state indicators under each functional module of the terminal are mined respectively and objectively. Finally, by combining the comprehensive weight of each indicator with the objective item and the fuzzy state evaluation method, the terminals' state evaluation completely based on the data is realized, and then the method will is verified with the reality example.

2. Partition and association rule mining of distribution terminal functional module

2.1. Functional module division of Distribution terminals

As complicated electronic equipment without standardized structure, distribution terminals' equipment health and functional integrity are coupled with the equipment, internal and external factors. In order to improve the standardization and generalization level of terminals' state evaluation process, the standardized functional structure and module division process of distribution terminals are firstly analyzed.

As the hardware structure and component composition of the terminals may vary depending on the model of the equipment and the manufacturer's process level, the diversity of the hardware structure will affect the standardized performance of the state evaluation method. Although there are hardware differences such as component structure and model of terminal equipment, the basic functions of its configuration have the requirements of standardization. The functional structure of the distribution terminal is divided into five modules, which are regarded as the second-order objects of terminals' state evaluation: CPU module, Power module, Communication module, Measure module and Control module, which are denoted as the set of terminals' functional modules: $M = [M_1, M_2, M_3, M_4, M_5]$.

The running state of distribution terminal is divided into four evaluation elements, which constitute an evaluation set: $E_i = [e_1, e_2, e_3, e_4]$, representing the five functional modules of the distribution terminals in turn, e_1, e_2, e_3, e_4 respectively representing the normal, attention, abnormal and serious state of each functional structure.

2.2. Preprocessing of state parameters

The state parameters related to the operation of the distribution terminals are extracted from the internal operating parameters, external communication interaction and environmental state. The parameter names and labels determined in this paper are shown in table 1.

The calculation formula of SOH (State of Health) of battery is shown in formula (1):

$$SOH = \frac{C}{C_N} \tag{1}$$

Where C_N is the rated capacity of the battery in the terminal; C is the current full charge of the battery.

Table 1. Table of terminal state parameter.

Parameter	Symbol	Parameter	Symbol	Parameter	Symbol
Time lag of GPS	$z1$	Secondary load of PT/CT	$z2$	Insulation resistance	$z3$
Secondary load of PT/CT	$z4$	Signal strength	$z5$	The off-line rate	$z6$
The number of maloperations	$z7$	Battery's SOH	$z8$	Period of message missing	$z9$
Telemetry deviation	$z10$	Harmonic content	$z11$	Number of message missing	$z12$
Time delay of communication	$z13$	Relative humidity	$z14$	Performance period	$z15$
Temperature	$z16$	Humidity	$z17$	A type terminal's fault rate	$z18$
B type terminal's fault rate	$z19$	C type terminal's fault rate	$z20$	D type terminal's fault rate	$z21$
The state of CPU module	$z22$	The state of measure module	$z23$	The state of control module	$z24$
The state of power module	$z25$	The state of communication module	$z26$		

The statistical calculation process of terminals' off-line rate is shown in formula (2). Its trigger is started by the total call message of the central station, which recorded every Δt minutes ,and the statistic times of the normal response within the observation period (unit: day) are calculated as N_{off} .

$$Rate_{off} = 1 - \frac{N_{ol} \times \Delta t}{24 \times 60 \times T} \tag{2}$$

In addition, in order to explore the influence of common family defects in distribution terminal units on state evaluation, the fault rate of four types of distribution terminal in the observation range is taken as the parameter to be analyzed, and the fault frequency of the same type of terminal is calculated from the historical data of terminal operation and maintenance under the same statistical conditions.

When the status of each terminal module is abnormal or fault, the specific function missing module confirmed by the test will be marked as 1 in case of abnormal or fault, and 0 in case of normal. In addition, other state data without special description are all default to be general signal.

According to the above selected state data, it can be divided into numerical type and Boolean type, in which the numerical data can be divided into larger and better type, smaller and better type, and interval optimal type according to the response results. They all have their own dimensions. In order to facilitate calculation and analysis, they are firstly standardized according to the data characteristics. Then, in order to facilitate subsequent mining and processing of association rules between state parameters, according to whether the difference between state parameters and standardized operating state is too large, the numerical parameter is changed to Boolean type, that is, 1 means significant deviation from standardized operating state, and 0 means within the tolerable variation range of parameter standards.

2.3. Association rules mining based on Apriori algorithm

With the improvement of smart grid information data collection and storage capacity, power distribution terminals have accumulated a large number of highly redundant parameter data in the process of operation and maintenance. The correlation between data parameters and terminals' functional state can be analyzed according to the data mining method.

Association Rules are used to formalize the causal dependency relationship between the front items X and back items Y, namely , it is expressed as $X \Rightarrow Y$, and then quantify the strength of the association relationship. Therefore the criterion of state evaluation of distribution terminal function module will be explored.

Y_i is the back items of the association rules, which is used to indicate whether the functional modules of the distribution terminals are abnormal, with an outlier value of 1 and a normal value of 0. Each data parameter recorded and collected in the running process of the terminal is marked as x_j ($j=1,2,...k$), then the possible combination of k data parameters constitutes items X_n ($n=1,2,...,2^k-1$) as the front ones of association rules.

Set D be the sample data set of parameters and functional module states, which accumulated in the historical operation and maintenance process of the distribution terminal. For mining association rules $X_n \Rightarrow Y_i$, generally, support and confidence degree are used to quantify the strength of association rules, and lift degree is used to judge the validity of association rules (The higher lift degree is, the higher correlation between the front items and back items is. In this paper, it is considered that the association rules with the degree of lift greater than 2.5 are valid), and their calculations are shown in equations (3), (4) and (5) respectively. In order to extract the association rules with high credibility as far as possible from the factors related to the state of terminals, the threshold of support degree was set as 0.01 in this paper, and the threshold of confidence degree was set as 0.4.

$$S(X_n \Rightarrow Y_i) = P(X_n \cup Y_i \mid D) = \frac{num(X_n \cup Y_i)}{num(D)} \tag{3}$$

$$C(X_n \Rightarrow Y_i) = P(Y_i \mid X_n) = \frac{num(X_n \cup Y_i)}{num(X_n)} \tag{4}$$

$$L(X_n \Rightarrow Y_i) = \frac{P(Y_i \mid X_n)}{P(Y_i)} = \frac{num(X_n \cup Y_i) \times num(D)}{num(X_n) \times num(Y_i)} \tag{5}$$

Where $num(\cdot)$ represents the counting operator that relies on the sample data of terminals under the corresponding set of items.

In order to realize the state evaluation of distribution terminals' modules, the Apriori algorithm adopted in this paper is constrained: the back item is limited to the state of a certain functional module of the distribution terminals, and the front item only considers the influence of single factor and does not explore the potential rules under their mutual combination, so as to reduce the search scope and time cost.

3. Fuzzy comprehensive evaluation of terminals' functional module

The result of association rule analysis is used as the indicators system of state evaluation of terminal module $R_i = \{r_{i,j} \mid j \in N\}$, where $r_{i,j}$ represents the validity evaluation indicators mined by association rules. Then, according to the fuzzy comprehensive evaluation methods, the status of the functional layer and the equipment layer of the terminal is evaluated.

3.1. The subjective and objective comprehensive weights of indicator sets

For the established indicator $r_{i,j}$, in order to quantify the influence degree and level of the difference on the overall state of the terminal, it is often necessary to assign weight according to its importance. In the traditional weighting method, equipment experts generally judge the importance of each indicator based on experience, that is, the analytic hierarchy process quantifies the weight. The process of determining subjective weights $\omega'_{i,j}$ originates from the equipment operation and maintenance experience of the judgment. To some extent, it lacks objectivity and is difficult to implement.

The objective weight derived from the analysis of data samples is determined by using the performance indicator in the mining process of association rules, and its calculation formula is as follows:

$$\omega''_{i,j} = \frac{C_{i,j} + L_{i,j}}{\sum_{j=1}^{N_i}(C_{i,j} + L_{i,j})} \tag{6}$$

Where $C_{i,j}$ and $L_{i,j}$ respectively represents the confidence and promotion degree corresponding to the indicators in the mining process of association rules.

Then, the calculation formula of subjective and objective comprehensive weight of indicator $r_{i,j}$ is shown in equation (7). For the M_i functional module state of the distribution terminal, the weight matrix of each indicator is $W_i = [\omega_1, \omega_2 ..., \omega_{N_i}]$.

$$\omega_{i,j} = \frac{\omega'_{i,j} + \omega''_{i,j}}{2} \tag{7}$$

3.2. Comprehensive fuzzy evaluation method

At present, the commonly used models of membership functions obey triangle distribution, half trapezoid distribution, S-type distribution, Gaussian distribution and etc. Considering that in the multi-indicators system, the construction and calculation process of triangle combined with half trapezoid membership function is relatively simple and accurate compared with the complex membership function, this paper chooses this method to calculate the membership degree of each indicator.

$$a_1(x) = \begin{cases} 1 & x \le X_1 \\ \dfrac{X_2 - x}{X_2 - X_1} & X_1 < x \le X_2 \\ 0 & x > X_2 \end{cases} \tag{8}$$

$$a_2(x) = \begin{cases} \dfrac{x - X_1}{X_2 - X_1} & X_1 < x \le X_2 \\ \dfrac{X_3 - x}{X_3 - X_2} & X_2 < x \le X_3 \\ 0 & else \end{cases} \tag{9}$$

$$a_3(x)=\begin{cases}\dfrac{x-X_2}{X_3-X_2} & X_2<x\le X_3\\ \dfrac{X_4-x}{X_4-X_3} & X_3<x\le X_4\\ 0 & else\end{cases} \tag{10}$$

$$a_4(x)=\begin{cases}0 & x\le X_3\\ \dfrac{x-X_3}{X_4-X_3} & X_3<x\le X_4\\ 1 & x>X_4\end{cases} \tag{11}$$

Where x represents the evaluation value of terminal deterioration according to the evaluation indicator, $a_i(x)$ is the membership function under each state evaluation.

According to the above mentioned membership function, the fuzzy relation matrix of the corresponding terminals corresponding to each functional module can be written as $(F_{M_i})_{m_i\times 4}$, where the subscript M_i represents the functional module label of the intelligent terminal.

According to equation (12), the fuzzy comprehensive evaluation of distribution terminal functional layer is calculated.

$$J_{M_i}=W_iF_{M_i} \tag{12}$$

Finally, the fuzzy comprehensive evaluation of the equipment layer, that is, the overall state of the terminal, is carried out by formula (13).

$$J=\left[\omega_{M_1},\omega_{M_2},\cdots,\omega_{M_5}\right]\begin{bmatrix}J_{M_1}\\ J_{M_2}\\ \vdots\\ J_{M_5}\end{bmatrix} \tag{13}$$

Where ω_{M_i} is the weight of terminal functional module; Results according to the principle of maximum membership, then the status evaluation results of distribution terminal are finally determined.

4. Example

In a power distribution terminal operation tag samples collected by the data sets (a total of 600 data, covering 26 state of relative terminal), for mining association rules of terminal state evaluation, and analysis of one distribution terminal with a fault has occurred, the state data before the fault is to be obtained, on the state evaluation, to verify that the method described in this paper.

4.1. Comprehensive fuzzy evaluation method

According to the Apriori algorithm mentioned above, the scatter diagram of 18 association rules mined from the data set is shown in figure 1. The mined rules have high confidence and lift degree, which proves their validity. The pattern diagram is shown in figure 2, in which the size of the circle represents confidence degree and the color depth represents lift degree. It can be seen that according to the constraints in the paper, the mined rules refer to the status of terminal's functional modules by the transaction front item, that is, the status evaluation factors of terminal's functional modules are determined based on the data.

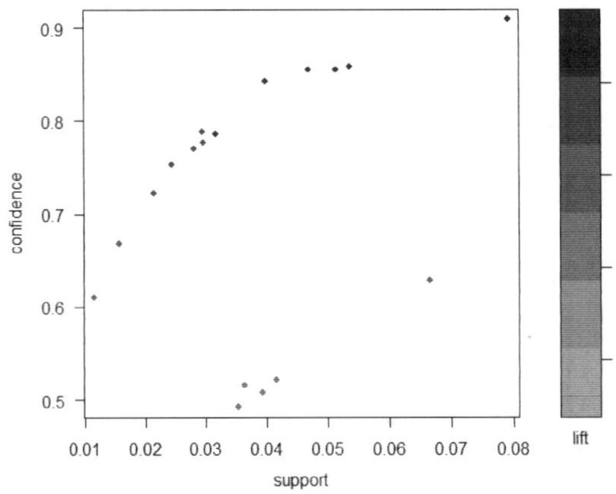

Figure 1. Scatter diagram of association rules.

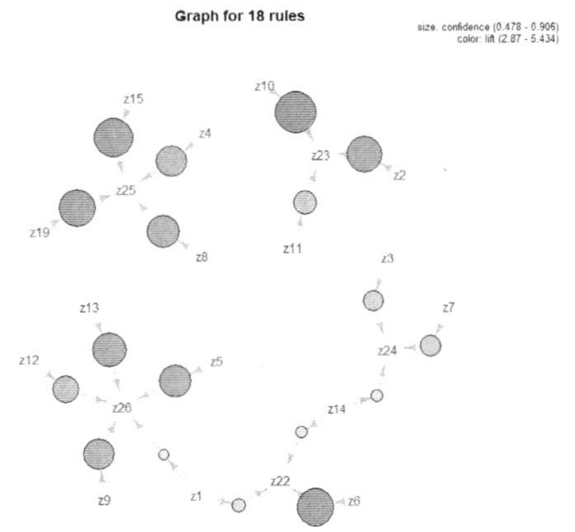

Figure 2. Pattern diagram of association rules.

Table 2. Association rule table of terminal state parameters.

Association rules	Support	Confidence	Lift
$z1 \Rightarrow z22$	0.040	0.522	3.130
$z6 \Rightarrow z22$	0.047	0.848	5.091
$z14 \Rightarrow z22$	0.037	0.500	3.000
$z2 \Rightarrow z23$	0.042	0.828	4.966
$z10 \Rightarrow z23$	0.080	0.906	5.434
$z11 \Rightarrow z23$	0.017	0.667	4.000
$z3 \Rightarrow z24$	0.013	0.615	3.692
$z7 \Rightarrow z24$	0.067	0.635	3.810
$z14 \Rightarrow z24$	0.037	0.500	3.000
$z4 \Rightarrow z25$	0.026	0.763	4.571
$z8 \Rightarrow z25$	0.030	0.783	4.696
$z15 \Rightarrow z25$	0.050	0.857	5.143
$z19 \Rightarrow z25$	0.053	0.842	5.053
$z1 \Rightarrow z26$	0.367	0.478	2.870
$z5 \Rightarrow z26$	0.030	0.783	4.696
$z9 \Rightarrow z26$	0.027	0.762	4.571
$z12 \Rightarrow z26$	0.020	0.706	4.235
$z13 \Rightarrow z26$	0.033	0.800	4.800

The mined results of association rules which shown in table 2, found that through the Apriori algorithm could not only mine general rules, such as $z12 \Rightarrow z26$,which means high frequent of unresponsive message is related to the abnormal state of communication module, but also could mine special rules which are hard to find in usual ways, such as $z19 \Rightarrow z25$, $z14 \Rightarrow z22$ and $z14 \Rightarrow z24$. The rule $z19 \Rightarrow z25$ finds that when the fault rate of B type terminal units' fault mostly happened in power module. The rules $z14 \Rightarrow z22$ and $z14 \Rightarrow z24$ find that the CPU module and control module are much more easy to drop into abnormal state when the relative humidity is rising high, which is well known that the high humidity environment might cause condensation at the electronic circuit part inside the terminal units. Then the subjective and objective comprehensive weight of each indicator is determined by the method described in this paper. The evaluation's indicator and weight of each functional layer of distribution terminal are shown in table 3.

Table 3. Terminal status indicators and weights table

Equipment layer	Functional layer	Indicators	Weights		
			ω'	ω''	ω
Terminal units	CPU module	z1	0.323	0.279	0.301
		z6	0.441	0.454	0.448
		z14	0.236	0.267	0.251
	Measure module	z2	0.573	0.345	0.459
		z10	0.326	0.377	0.352
		z11	0.101	0.278	0.189
	Control module	z3	0.327	0.352	0.340
		z7	0.421	0.363	0.392
		z14	0.252	0.285	0.268
	Power module	z4	0.367	0.235	0.301
		z8	0.301	0.241	0.271
		z15	0.237	0.264	0.250
		z19	0.095	0.260	0.178
	Communication module	z1	0.131	0.136	0.134
		z5	0.167	0.221	0.194
		z9	0.257	0.216	0.236
		z12	0.241	0.200	0.220
		z13	0.204	0.227	0.216

4.2. Stratified state evaluation of terminal units

According to the method described in this paper, the state evaluation of a distribution terminal is carried out, and the degradation degree of the terminal corresponding to each indicator is substituted into equation (8)-(11) to obtain the fuzzy relation matrix of each functional module, and then the fuzzy comprehensive evaluation result on the functional layer is obtained from equation (12) as follows:

Fuzzy relation matrix of module CPU module:

$$J_{M_1} = [0, 0.3753, 0.3831, 0.2416]$$

Fuzzy relation matrix of module measure module:

$$J_{M_2} = [0.5616, 0.3313, 0.1071, 0]$$

Fuzzy relation matrix of control module:

$$J_{M_3} = [0.6423, 0.2317, 0.1260, 0]$$

Fuzzy relation matrix of power module:

$$J_{M_4} = [0.2371, 0.3682, 0.3947, 0]$$

Fuzzy relation matrix of communication module:

$$J_{M_5} = [0, 0.2812, 0.5643, 0.1545]$$

Finally, the overall state evaluation result of the terminal on the device layer is obtained: $J = [0.1253, 0.2217, 0.3489, 0.3041]$.

Finally, the overall status evaluation result of the terminal on the equipment layer is obtained: that is, the result of status evaluation is found to be abnormal from the data of the terminal one week before the fault, which means it is in the transition period before the fault. The state history of the terminal indicates that the communication link intermittently initialized before the terminal was completely out of service. It proved that the state evaluation method in a certain data could indicate terminal's actual

IOP Conf. Series: Materials Science and Engineering **752** (2020) 012015 doi:10.1088/1757-899X/752/1/012015

operation state, and according to the results of state evaluation, it should as soon as possible to arrange the relative maintenance and replacement of the abnormal terminal, in order to shorten the time period without monitoring and measurement for distribution network influenced by the fault terminal.

5. Conclusion

This paper presents a state evaluation method of intelligent distribution terminals based on association rules. Firstly, the Apriori algorithm is used to mine the association rules of each functional module from the multi-dimensional state parameters and then establish the corresponding state evaluation's indicators, using confident degree and lift degree to determine the objective weight of each indicator. Finally, combined with the fuzzy comprehensive evaluation, the state evaluation on the functional layer and the equipment layer of the terminal was carried out successively, and the following conclusions were obtained:

1) Compared with the general fuzzy comprehensive evaluation method, whose indicators establishment is completely derived from summary experience, the method proposed in this paper establishes the indicators system based on data analysis, which greatly increases the objectivity and validity of state evaluation method for terminals.

2) In addition, the indicators system can change according to the update and expansion of the data set, making the method more suitable for the situation of state evaluation of terminal equipment with fuzzy environment or high data quality.

Acknowledgments

This work was supported by Science & Technology Project of Guangdong Power Grid Co. Ltd. (GDKJXM20172867)

References

[1] Cong W, Sheng Y R, and Qi G F 2018 *J. Automation of Electric Power Systems.* **42(15)** 77-85
[2] Zhang Z H, Zhou J, and Cai Y M 2017 *J. Automation of Electric Power Systems.* **41(13)** 106-110
[3] Zhang J C, Chen L, and Zhang M Y 2019 *J. High Voltage Engineering.* **45(6)** 1729-1736
[4] Zhang Y Q, Kou L F, and Sheng W X 2016 *J. Power System Technology.* **40(3)** 768-773
[5] Li G, Zhang B, and Zhao W Q 2016 *J. Automation of Electric Power Systems.* **44(6)** 31-35
[6] Ma Z, Zhou L M, Yuan H W 2019 *J. Proceedings of the CSEE.* **39(1)** 130-140
[7] Jiang X C, and Sheng G H 2018 *J. High Voltage Engineering.* **44(4)** 1041-1050
[8] Xie C Z, Bai J F, and Wang H B 2018 *J. Proceedings of the CSEE.* **38(21)** 6233-6244

IOP Conf. Series: Materials Science and Engineering **752** (2020) 012016 doi:10.1088/1757-899X/752/1/012016

Fault Diagnosis of Distribution Terminal Units' Measurement System Based on Generative Adversarial Network Combined with Convolutional Neural Network

Q F Zhou[1], Y Ge[1], X D Song[2,*], K Lai[3] and J C Guo[4]

[1]Guangdong Power Grid Co. Ltd., Guangzhou, Guangdong 510000, China
[2]Guangdong Diankeyuan Energy Technology Co. Ltd., Guangzhou, Guangdong 510080, China
[3]Jiangmen Power Supply Bureau of Guangdong Power Grid Co. Ltd., Jiangmen, Guangdong 529000, China
[4]School of Electrical Engineering, Southeast University, Nanjing, Jiangsu 210096, China

[*]694011613@qq.com

Abstract. As the condition monitoring and control device in the distribution automation system, the abnormal or fault state of distribution terminal units' measurement system will negatively affect the quality of measured electrical quantities, and therefore, the fast and accurate discrimination of the abnormal state's data will improve the reliability of distribution automation system. This paper proposes a method, which is based on generative adversarial network (GAN) combined with convolutional neural network (CNN), to discriminate the specific fault category of distribution terminals' measuring electrical data. Firstly, four fault state characteristic time-frequency domain graph models of terminals' AC voltage sampling data are established, which based on Fourier transform (STFT). Then, take advantage of GAN's reconstruction of input graph data, to generate additional time-frequency sample graphs expanding the sample size of training set, which will be used to train a CNN to diagnosis and classify the fault state of terminals from measuring data. Finally, three training sets with different capacity expansion modes are set up to compare and verify that the method of GAN combined with CNN proposed in this paper improves the discrimination accuracy on the fault data, and the validation of diagnosis terminals' measurement system.

1. Introduction

The distribution terminal units, which usually installed in the medium voltage substation, are the important condition monitoring and control equipment in distribution automation system [1-3]. Affected by the installation environment, different product quality and performance factors, the terminals' measurement system frequently dysfunction or abnormally operate, for these relative problems, however, lacking online fault diagnosis and discriminant methods make the negative effect of terminals' fault much worse [3]. Therefore, it is necessary to study a method to fully exploit the information potential contained in the measured electrical data of terminals, and to diagnosis measurement system's fault by judging whether the electrical monitoring data state of the distribution terminal are normal, so as to improve the reliability of the terminals.

Content from this work may be used under the terms of the Creative Commons Attribution 3.0 licence. Any further distribution of this work must maintain attribution to the author(s) and the title of the work, journal citation and DOI.
Published under licence by IOP Publishing Ltd

As for the research on discrimination method of electrical equipment's abnormal feature data, literature [4] established the BP neural network to complete identification fault type based on some typical faults sampled data of voltage transformer, however, the identification accuracy and generalization performance of the method still have great room to improvement. Literature [5] used Elman neural network's strong adaptability to diagnose the fault of high-voltage measurement system, but the capacity of analysis feature samples was not enough, which might affect the results of fault diagnosis and discrimination. Literature [6-7] respectively used the method of Wavelet Transformation to extract feature of fault and comparison of transverse and longitudinal current to identify the fault of the transformer. However, the selection of fault features was subjective and incomplete to a certain extent. Literature [8] deeply analysed mechanism and equivalent model of harmonic type fault for transformer, but did not conduct any in-depth research on discrimination based on data analysis.

For the research on deep learning methods in the aspects of learning data feature and identification of data pattern, literature [9] realized the accurate classification of time-frequency diagrams of different bearing vibration faults through deep CNN. Literature [10] studied Wavelet Transform method to generate wavelet map for the vibration signals of loose transformer windings and loose iron cores, and generate grey-scale feature graph, which were used to train CNN, and then successfully realized identification of the vibration signals the deep net. The above studies are all based on CNN and have achieved good results in various pattern recognition and classification problems. However, the sample size of training required by the CNN is large, and the over-fitting problem of the model on the small-scale training sample has not been solved yet. For the enhancement and repair of feature data, literature [11] trained GAN to learn the ability of generating new face pictures, which improved its great benefit on data refactoring and enhancement.

This paper proposes a deep learning method based on GAN combined with CNN, aiming to accurate discrimination abnormal voltage measured by distribution terminals' measurement system. Firstly, four fault state characteristic time-frequency domain graph models of terminals' AC voltage sampling data are established, which based on STFT. Then, take advantage of GAN's reconstruction of input graph data, to generate additional time-frequency sample graphs expanding the sample size of training set, which will be used to train a CNN to diagnosis and classify the fault state of terminals from measuring data. Finally, three training sets with different capacity expansion modes are set up to compare and verify that the method of GAN combined with CNN proposed in this paper improves the discrimination accuracy on the fault data, and the validation of diagnosis terminals' measurement system.

2. Fault feature model of voltage data

According to the characteristics of ac voltage between the normal and abnormal state of the distribution terminals' voltage measurement and acquisition system, the four feature data models are respectively established, as the object of data discrimination and fault diagnosis.

2.1. Measurement data model of normal state

Due to the interference from environment, the distribution terminals' measured voltage data might contain a certain amount of random noise in actual operation. Generally, it can be considered that the noise part of data is subject to Gaussian distribution. The normal state data model considering the error of acquisition and measurement is shown in the following equation:

$$f(\mathrm{t}) = A\sin(\omega \mathrm{t} + \varphi) + N(0, \delta^2) \quad (1)$$

Where, $N(0,\delta^2)$ is the measurement noise of Gaussian distribution with mean of zero and variance of δ^2, in turn, A, ω and φ are the amplitude, angular frequency and phase angle of the measured voltage data.

2.2. Measurement data model of accuracy distortion

When the fault occurs in the signal processing module such as the transformer and signal transmission unit in the distribution terminals' measurement system, the accuracy of the measured signals will be distorted, and the noise part may has the feature of marked change. In the case of precision distortion fault, the average value of voltage measurements remains unchanged and the measurement variance changes, expressed as $N_f(0,\delta_f^2)$. The fault data model at this time is shown in the following equation:

$$f(t) = A\sin(\omega t+\varphi) + N(0,\delta^2) + N_f(0,\delta_f^2) \quad (2)$$

2.3. Measurement data model of decaying oscillation

When the measurement device in the distribution terminals is aging or decaying, the measured data usually have a certain characteristic of oscillation attenuation. The fault data model at this time is shown in the following equation:

$$f(t) = e^{-b\cdot t}A\sin(\omega t+\varphi) + N(0,\delta^2) \quad (3)$$

Where, $e^{-b\cdot t}$ is exponential attenuation coefficient; b is a real number greater than 0, which expressing the magnitude of exponential decay; $N(0,\delta^2)$ is the Gaussian noise term of such measurement data, indicating the measurement error.

2.4. Measurement data model with harmonics

The measurement element in the distribution terminal may deviate from the fundamental frequency due to environmental interference or component fault. For example, electromagnetic transformer will occur when the excitation characteristic of electromagnetic transformer is degraded, resulting in the measurement containing high-order harmonic components [8]. The fault data model at this time is shown in the following equation:

$$f(t) = A\sin(\omega t+\varphi) + A_2\sin[2\omega t+\varphi_2] + A_3\sin[3\omega t+\varphi_3] + \ldots + N(0,\delta^2) \quad (4)$$

2.5. Time-frequency graph of the measurement data model

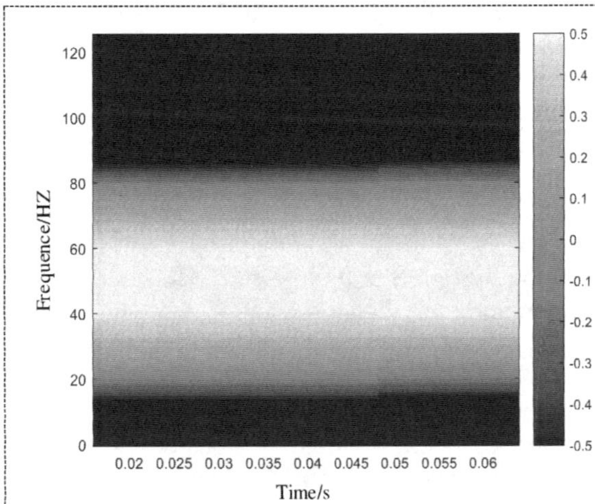

Figure 1. Time-frequency graph of normal-state data.

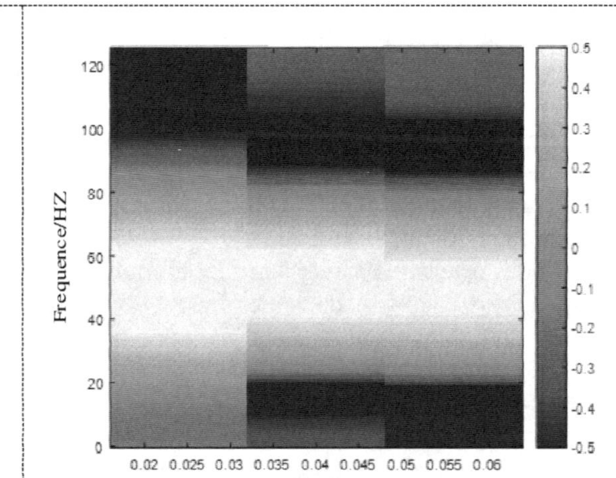

Figure 2. Time-frequency graph of accuracy distortion data.

IOP Conf. Series: Materials Science and Engineering **752** (2020) 012016 doi:10.1088/1757-899X/752/1/012016

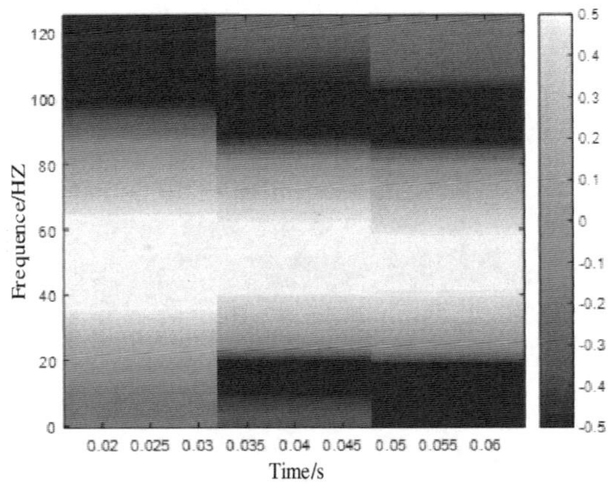

Figure 3. Time-frequency graph of decaying oscillation data.

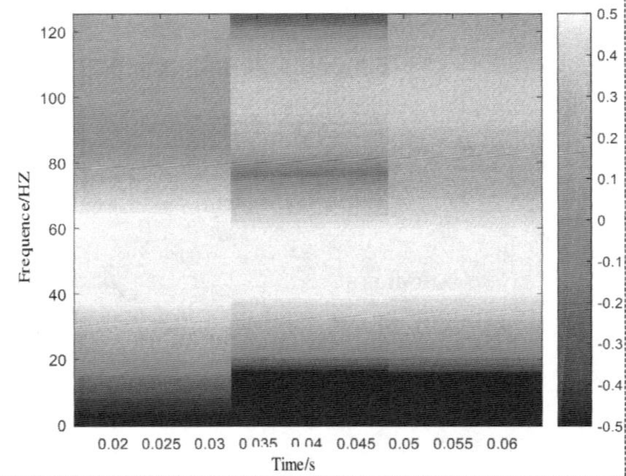

Figure 4. Time-frequency graph of harmonics data.

The data models of the distribution terminal established above belong to time-domain. In order to strengthen the classification characteristics of different data models, STFT is introduced to convert the digital data samples into time-frequency graph containing time-domain and frequency-domain feature information.

Based on the result of the STFT, the measurement data of different time-frequency graphs, as shown in figure 1-4, horizontal axis represents time sampling data and the vertical axis represents the different frequency components, and the depth of the graph's pixel colour represents the amplitude of different frequency component at the same time. Generally, the four different type time-frequency graphs are evidently different in the graphics characteristics such as shape, colour, therefore, it can be feed into CNN, converting the recognition of time-domain data to discrimination the feature pictures.

3. Discrimination of feature graph based on GAN combined with CNN

In order to give full play to GAN's ability of feature learning and data regeneration and CNN's huge advantages in image recognition. The identification and classification of fault feature data model, which based on GAN combined with CNN, comprehensively established in this paper is shown in figure 5. STFT is used to convert the AC voltage sample data into time-frequency feature graph samples, and the training set composed of time-frequency graphs of various types of faults feed into GAN to generate reconstructed time-frequency graphs, so as to expand capacity of four types of faults' training sets. Then, the expanded training sets are used to train CNN, so as to realize the classification and identification of four types fault feature data.

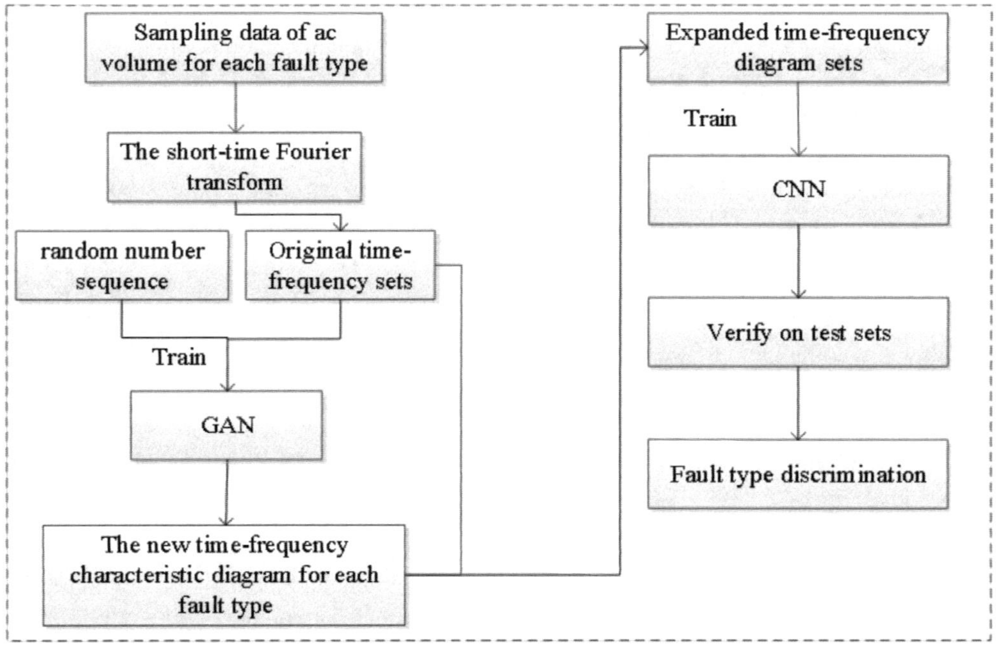

Figure 5. Fault identification flow chart of GAN and CNN.

3.1. Capacity expansion model of time-frequency graph based on GAN

GAN (Generative Adversarial Network) is a kind of probability generation model, which consists of the Generator G (Generator) and Discriminant D (Discriminator). It could learn the distribution of the original data through confrontation training.

The purpose of trained G is to imitate and master the distribution of sample data, making the performance of distribution $G(z)$ on D, which expressed as $D(G(z))$, as consistent as possible with the performance of real data x on D, which expressed as $D(x)$, so that D cannot distinguish the generated samples from real samples. The purpose of trained D is to correctly distinguish fake samples from generated samples. The above logical relationship can be expressed as maximum-minimum problem [11], and its mathematical expression is as follows:

$$\min_G \max_D \underset{x\sim p_r}{E}[\log(D(x))] + \underset{x\sim p_g}{E}[\log(1-D(z))] \tag{5}$$

Where, x is characteristic sequence and z is random number sequence; p_r represents the time-frequency feature graphs of the original true data; p_g represents the time-frequency feature graphs of new fake data generated by G.

D and G, through continuous cross optimization, improve their respective discrimination and generation ability respectively. This optimization process is to find the Nash equilibrium of the game process between them. When the training stops, the discriminator D could hardly distinguish the difference between the time-frequency graphs of distribution terminals' fault feature data, which generated by the generator, and the time-frequency diagram of the original data.

In order to facilitate the processing of time-frequency graphs of image types, deep convolutional generative adversarial networks (DC-GAN) are used in this paper, which is composed of two mirrored convolutional neural networks. In addition to the output layer of G and the input layer of D, batch normalization is added to all other layers, which will help stabilize the optimization process of objective function gradient to some extent.

3.2. Time - frequency graph recognition model based on CNN

Convolutional Neural Network (CNN) is a multilayer perceptron designed to identify two-dimensional feature graphs. It is a deep network model with multiple hidden layers, which can transform low-level features into high-level features through feature transmission layer by layer, so as to realize feature learning and expression. Compared with BP neural network and other shallow networks, CNN has a stronger ability to learn and express complex features, and a faster operation speed, and could avoid the problem of getting into the local extreme.

The time-frequency graphs of various data samples are feed into CNN, and then after process of pre-processing and normalization, the input graphs are successively processed through convolution layer, pooling layer and the full connection layer, and the feature extraction and dimensionality reduction of data are completed through activation function, and finally output the type number of fault feature data, which express the identification and classification of fault feature data.

As for the convolution layer of CNN, different convolution kernels are used to convolve with the feature graph, and the feature graph after dimensionality reduction is obtained by activating the function. The convolution process could be expressed as follows:

$$x_j^l = f\left(\sum_{i\in M_j} x_j^{l-1} * k_{ij}^{l-1} + b_j^l\right) \tag{6}$$

Where: x_j^l represents the output j of the layer l ; k_{ij}^l represents a convolution kernel of the convolution layer; $*$ represents the convolution operation; b_j^l represents the bias of the convolutional layer; $f(\square)$ is an activation function, and the ReLU function is commonly used one, which could expressed as follows:

$$f(x) = \begin{cases} x & x > 0 \\ 0 & x \le 0 \end{cases} \tag{7}$$

In the pooling layer, the maximum value (maximum pooling) or average value (average pooling) of the feature graph output for the convolution layer in each non-overlapping region of size $n\times n$ is selected, so as to further reduce the dimension of the feature graph.

In the full connection layer, input all the one-dimensional feature vectors expanded by the feature graph of the previous layer, and output the final classification after weighted and imported into the activation function. In this paper, the Softmax function is used as the activation function of output layer.

Based on the classification task implemented by CNN, the common loss functions include mean square error function, cross entropy function and negative logarithmic likelihood function. The cross entropy function which has good performance is selected in this paper, and the expression is as follows:

$$loss = -\frac{1}{n}\sum_{i=1}^{n}[y_i \ln \hat{y}_i + (1-y_i)\ln(1-\hat{y}_i)] \tag{8}$$

Where, n is the number of fault feature samples; y is the true value; $\hat{y}$ is the predicted value.

4. Example

4.1. Power distribution terminal ac sampling data sample generation

By the function (1) - (4) representing the distribution of terminals' fault voltage feature sampling data samples, which were generated and simulated in Matlab, the set for sampling frequency of AC voltage is 250 HZ, and sampled four cycles of power frequency measurement data points (20 voltage measurement points) composed as one data sample. Adjust the parameters of each data model for simulation, as seen in table 1 for specific simulation parameters and sample size.

Table 1. Simulation parameter setting table of voltage model

Type	Parameter		The number of samples
Normal	$\varphi \subset U[0,2\pi]$ $\delta \subset U[0,0.1]$		300
Accuracy distortion	$\varphi \subset U[0,2\pi]$ $\delta \subset U[0,0.1]$ $\delta_f \subset U[0.1,0.2]$		300
Damped oscillation	$b \subset U[5,20]$ $\delta \subset U[0,0.1]$ $\varphi \subset U[0,2\pi]$	$A=1.2$ $t_n \in [0,0.08]$ $n=1,2,...,20$	300
Harmonic interference	$A_2 \subset U[0,A]$ $\varphi_2 \subset U[0,2\pi]$ $A_3 \subset U[0,A]$ $\varphi_3 \subset U[0,2\pi]$		300

Operating Matlab, use spectrogram function, which is a short-term Fourier analysis function, to convert the simulated sampled voltage data into time-frequency graph samples.

4.2. Combined neural network structure and fault identification results analysis

Table 2. Structure parameters table of DCGAN

Component	Structure	Dimension of input [height, width, depth]	Dimension of output [height, width, depth]
D	Con Layer1	[64,64,3]	[32,32,64]
	Con Layer2	[32,32,64]	[16,16,128]
	Con Layer3	[16,16,128]	[8,8,256]
	Con Layer4	[8,8,256]	[4,4,512]
	Full-connected Layer5	[1024]	[1]
G	Full-connected Layer6	[100]	[4,4,1024]
	De-con Layer7	[4,4,1024]	[8,8,12]
	De-con Layer8	[8,8,512]	[16,16,256]
	De-con Layer9	[16,16,256]	[32,32,128]
	De-con Layer10	[32,32,128]	[64,64,3]

Table 3. Structure parameters table of CNN

Structure	Dimension of input [height, width]	Numbers of kernels (or output dimension)	Dimension of Kernel [height, width]
Con Layer1	[64,64]	16	[3,3]
Pooling Layer2	[32,32,64]	16	[3,3]
Con Layer3	[16,16,128]	32	[3,3]
Pooling Layer4	[8,8,256]	64	[3,3]
Full-connected Layer5	[1024]	[1024]	[3,3]
Full-connected Layer6	[100]	[4]	[3,3]

Based on Tensorflow, which is the deep learning framework of Google, the processing of data time-frequency graph clipping and data standardization is carried out. After that, the network architectures of GAN and CNN were built and corresponding model parameters were set, and corresponding parameters are shown in table 2 and table 3.

This paper specifically selected LeNet-5 as CNN network structure, and Adam algorithm is adopted to optimize the loss function. The global learning rate is set to 0.0001, and each batch is 20 samples.

Against the new time-frequency graphs of all kinds of data samples generated by generator G in neural network, take normal data and attenuated oscillation data as examples. After training GAN with different periods, new time-frequency graph samples of data are generated, as shown in figure 6.With the increase of training step length, the time-frequency graph generated by GAN is closer to the real training sample, that is, the more features of simulation data samples are learned, so that the random sequence can generate new time-frequency graph data samples after passing through multiple trained de-convolution layers, so as to expand the sample size of the training set.

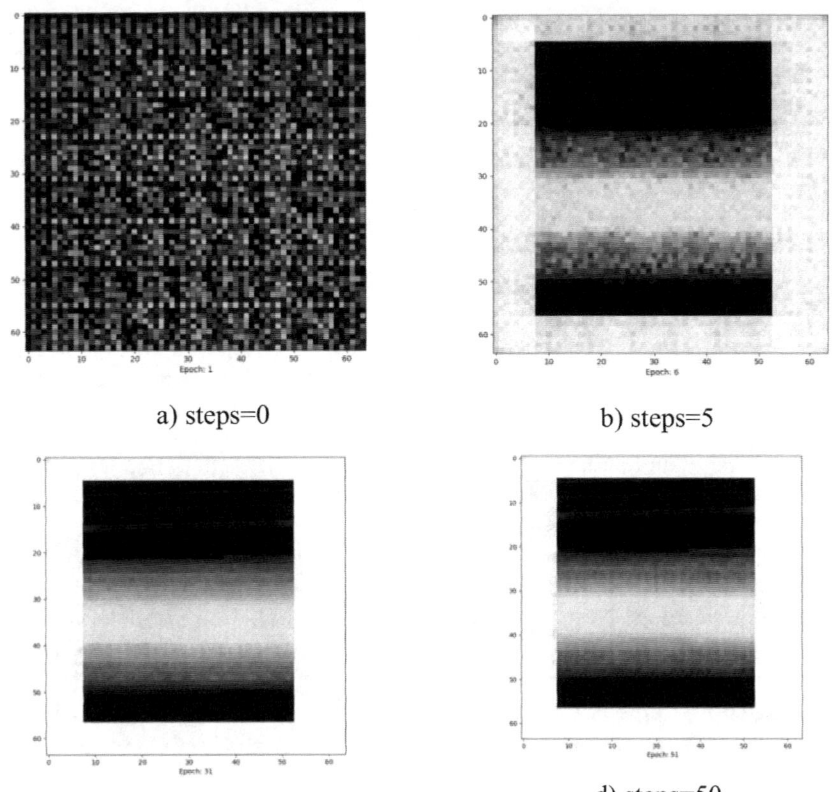

a) steps=0 b) steps=5

c) steps=30 d) steps=50

Figure 6. Time-frequency graph of normal samples generated based on GAN.

GAN (step=50) after training the training set generates 100 new time-frequency graphs of data samples of various types to expand the capacity of training set input into CNN network.

Four hundred time-frequency graphs were randomly selected from the simulation-generated data sample set as the test set of CNN fault classifier. Three different CNN training sets are set to compare the classification performance results of the methods used in this paper:

1) Training set A: all kinds of real data samples generated by the original simulation (without GAN expansion) are composed of $4 \times 200 = 800$ time-frequency characteristic graph.

2) Training set B: the mixed training set generated by GAN after expansion is composed of time-frequency characteristic graph of $4\times(200+100)=1200$, which contains both real data samples and new data samples generated by GAN.

3) Training set C: in the real data sample set generated by the original simulation (the capacity is 800), 400 time-frequency graphs samples are randomly copied to form the expanded training set, which is composed of $4\times(200+100)=1200$ time-frequency graphs. The expansion mode is only random replication on the original sample set, and actually no data samples with new characteristics are added. The purpose of setting the training set is to eliminate the influence of the training set capacity on the CNN fault classifier so as to better compare the performance results of the method used in this paper.

Different training sets were input into the CNN network, and the loss function values on the training set with different training steps were counted and compared with the test accuracy on the test set. The results were shown in figure 7 and figure 8 respectively.

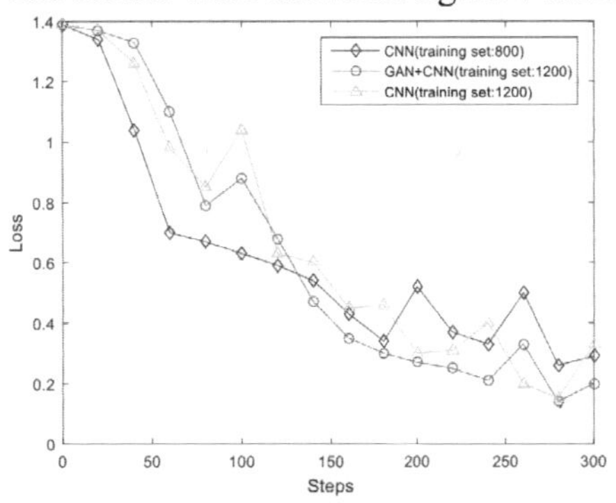

Figure 7. Loss value diagram of CNN training process.

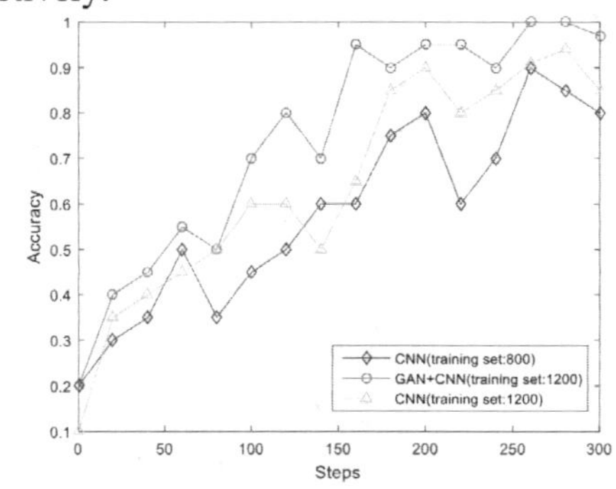

Figure 8. The recognition accuracy graph of CNN on test set.

5. Conclusion

This paper presents a fault diagnosis method of voltage sampling module of distribution terminals with GAN and CNN. Firstly, this paper established normal state voltage sampling data model containing noise part and three types of approximate mathematical model of abnormal state data, then through STFT converting simulation data into a comprehensive time-frequency graphs, and the generated data samples by 2:1 is divided into the original training set and testing set. On the basis of GAN, feature learning, pattern generation and expansion were carried out for the original training set of various types of data samples. Then, the expanded samples were used to train the CNN (lenet-5) model, and the performance results of fault classification on the same test set were obtained through three training sets with different capacities and expansion modes. The simulation data examples and the training verification results of combined neural network showed that:

1) The time-frequency graphs obtained by STFT could be much easier to distinguish the sampling data of distribution terminals with different fault types.

2) After GAN generated new data samples, the training sets B had a good classification result on the testing sets, indicating that GAN could learn and regenerate the features of limited data samples and realize the expansion of a limited number of sample data.

3) In the training process of CNN with the simulation data of time-frequency graph containing noise, the input values of CNN loss function of different capacity and different expansion mode training sets have little difference, and they all have a good decline process. The fault classification model based on CNN would have a good difference effect on various types of fault data.

4) The accuracy of training sets with different capacities and expansion modes is different on the same test set. Compared with the original training data set without expansion and the expansion data set after simple random replication, the accuracy of identification of the expansion training set

generated by GAN is significantly improved on the test set. It is proved that GAN combined with CNN can improve the generalization performance of fault time-frequency graph recognition model, namely, the recognition accuracy on unknown data. However, the specific mathematical relationship between GAN expansion capacity and the improvement of recognition accuracy on the CNN test set and the possible upper limit have not been proved in this paper, and further studies are still needed.

Acknowledgments
This work was supported by Science & Technology Project of Guangdong Power Grid Co. Ltd. (GDKJXM20172867)

References

[1] Zhang P, SU Y F and Han Y B 2018 *J. Proceedings of the CSU-EPSA.* **30(2)** 61-65
[2] Gong F L, Mu L H and Guo W M 2018 *J. Power System Protection and Control.* **46(1)** 30-36
[3] Luo F Z, Yang W T, and Zhang T Y 2018 *J. Automation of Electric Power Systems.* **42(19)** 77-85
[4] Jiang Y Y, Wang J L and Li J Y 2018 *J. Proceedings of the CSU-EPSA.* **30(6)** 134-139
[5] Liu Y Y, Le J and Li Q L 2017 *J. Electric Power Automation Equipment.* **37(3)** 167-174
[6] Lin G, W B and Chen Q J 2018 *J. Proceedings of the CSEE.* **38(11)** 3393-3401
[7] Wei D, Gong Q W and Lai W Q 2016 *J. Proceedings of the CSEE.* **36(S1)** 3393-3401
[8] Tan B D, Yang J and Lai Q P 2019 *J. Automation of Electric Power Systems.* **43(1)** 149-157
[9] Lin J H, Zhang Y and Zhao T 2019 *J. Proceedings of the CSEE.* **39(1)** 84-96
[10] Huang X B, Hu X W and Zhu Y C 2018 *J. Electric Power Automation Equipment.* **38(5)** 136-140
[11] Wang J R, Li S M and Han B K 2018 *J. IEEE Access.* **7** 168-180

IOP Conf. Series: Materials Science and Engineering **752** (2020) 012017 doi:10.1088/1757-899X/752/1/012017

Research on Parameter Identification Method of Generator Excitation System Based on Differential Evolution Algorithm

Nan Li[1,2,a], Yuqun Liu[2,b] and Bin Geng[3,c]

[1] Key Laboratory of Modern Power System Simulation and Control and Renewable Energy Technology, Ministry of Education, Northeast Electric Power University, Jilin 132012, Jilin Province, China
[2] College of Electrical Engineering, Northeast Electric Power University, Jilin 13012, Jilin Province, China
[3]Jilin Electric Power Co., Ltd., Changchun 130021, Jilin Province, China

[a]jllinan@163.com; [b]1343572496@qq.com; [c]672108470@qq.com

Abstract. In order to solve the problem of large parameter identification error caused by non-linear links of excitation system being triggered easily when transient stability is under fault state, an improved differential evolution algorithm for system parameter identification is proposed by using the characteristic of artificial intelligence algorithm that the nonlinear link is approximated infinitely through optimization. The improvement of the algorithm solves the problems of slow convergence speed, poor fine optimization ability and easily to produce local optimum when classical artificial intelligence algorithm identifies the parameters of non-linear links. At the same time, in order to solve the problem of inaccurate parameters in the whole identification, a decomposition link identification strategy is proposed. The example analysis shows that the algorithm improves the convergence speed, avoids local optimum and improves the convergence accuracy. According to the proposed parameter identification strategy, the excitation system is decomposed and identified, which improves the accuracy of generator excitation system parameter identification, and provides an accurate model and method for power system stability analysis

1. Introduction

In recent years, many scholars have done a lot of research work in the field of parameter identification of excitation system in [1]. Literature [2] uses genetic algorithm to identify the non-linear link of excitation system, which proves the feasibility of using artificial intelligence algorithm to identify the parameters of excitation system. In [3], the improved genetic algorithm is used to identify the non-linear excitation system, and the identification accuracy is improved by combining the genetic algorithm with the gradient descent BP neural network. In [4], the adaptive particle swarm optimization (APSO) algorithm is applied to identify the parameters of excitation system to improve the convergence speed. But there are many problems in practical application. For example, genetic algorithm, particle swarm optimization and other algorithms have many control parameters, complex implementation and slow convergence speed.

Content from this work may be used under the terms of the Creative Commons Attribution 3.0 licence. Any further distribution of this work must maintain attribution to the author(s) and the title of the work, journal citation and DOI.
Published under licence by IOP Publishing Ltd

Differential evolution algorithm has the characteristics of less control parameters and strong robustness. It has a good effect on muti-parameter optimization problem. But it also has the problems of slow convergence, local optimum and poor ability of fine optimization. In order to improve the convergence speed and the ability of fine optimization, the classical differential evolution algorithm is improved and applied to parameter identification of generator excitation system. In view of the high order of generator excitation system and the large number of parameters to be identified, a strategy to improve the accuracy of parameter identification based on link decomposition is proposed. According to the actual observation points and parameter sensitivity, the decomposition identification results show that the accuracy of parameter identification is improved.

2. Excitation System Model

We use self-shunt static excitation system model for parameter identification in [5], and the type model is FV. The block diagram of its transfer function is shown in Figure 1.

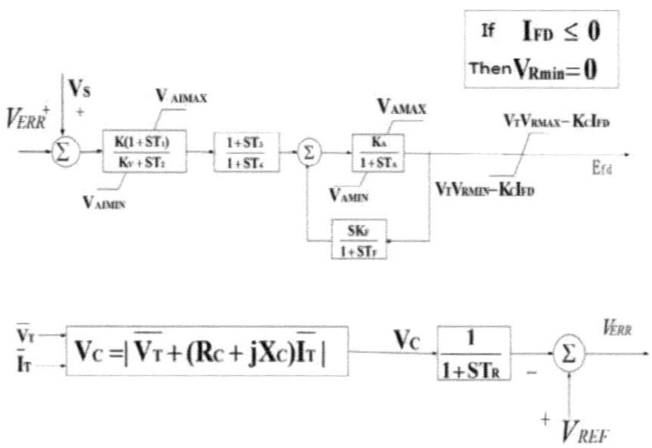

Figure. 1. Model of FV type self-shunt excitation system

3. Parameter identification based on improved differential evolution algorithm

Differential Evolution (DE) was proposed by Price and Storm in 1995. It is an evolutionary algorithm based on population parallel global random search. The basic steps of the algorithm are initialization, mutation, crossover and selection according to literature [6].

3.1. Improved mutation operation

3.1.1. For current non-optimal individuals. The weight coefficient λ is introduced, as shown in formula (1):

$$v_i(t+1)=x_{r3}(t)+F\times(\lambda\times(x_{r1}(t)-x_{r2}(t))+(1-\lambda)\times(x(t)_{best}-x_{r3}(t))) \tag{1}$$

Where, $\lambda=e^{1-GM(GM+1-G)}$, GM is the maximum number of iterations, $x(t)_{best}$ is the best individual in the current population,and r_1, r_2, r_3 are three different index numbers that are not equal to i. t is the current number of iterations, and F is the scaling factor. The optimization effect is better when the value is between [0.4, 1]. The larger the value of F, the faster the convergence speed, the smaller the value of F, the narrower the search range, and the higher the identification accuracy. During the operation of the algorithm, both convergence speed and identification accuracy need to be taken into account. Therefore, the random oscillation value with 0.6 as the center and 0.4 as the amplitude is chosen, as shown in formula (2):

$$F=0.6+(-1)^k\times0.4\times rand \tag{2}$$

Where, k is a random positive integer.

3.1.2. For the Current Optimal Individual. For the current population optimization, small step oscillation fine-tuning mutation is carried out to enhance the local optimization ability. The specific operation is shown as follows:

$$v_i(t+1) = x(t)_{best} + (-1)^k \times (x_i^U - x_i^L)/100 \tag{3}$$

Where, k is a random positive integer, and x_i^U and x_i^L are upper and lower bounds of parameters.

3.2. Improved mutation operation Improving cross-over operation

In this paper, the difference vectors between the individual and the optimal individual of the current population are introduced, so that the evolutionary direction of the population tends to the optimal individual of the current population. This paper takes CR as the center and 0.6 as the amplitude, taking into account the evolution speed and the diversity of the population. As shown in formula (4):

$$CR = 0.6 + (-1)^k \times 0.3 \times rand \tag{4}$$

4. Parameter identification based on improved differential evolution algorithm

Artificial intelligence algorithms such as Differential Evolution Algorithms can identify the model parameters as a whole, that is, the values of all the parameters to be identified at one time. Let the transfer function of the excitation system be:

$$G(s) = \frac{a_0 + a_1 s + \cdots + a_m s^m}{1 + b_1 s + \cdots + b_n s^n} \tag{5}$$

Where, a_0, a_1, $\cdots$, a_m, b_0, b_1, $\cdots$, b_n are the parameters to be identified.

Because the excitation system is a high-order system, there are many sets of parameters to be identified. Even in accordance with the general practice of engineering, using multiple identification to obtain the average value, there will still be large errors, and the correct results can not be obtained. In this paper, a parameter identification strategy based on link decomposition is proposed. The system is decomposed into several sub-modules to reduce the order of the system and improve the accuracy of parameter identification.

The parameter identification strategy of link decomposition is as follows:

- For both input and output signals, a simple sub-module with clear observation points can be established and identified directly.
- For modules with nested relationships, the parameters of simple links with clear structure and close distance to the observation point are identified in order from small to large and from inside to outside.
- The parameters with high error sensitivity are identified first, and then the parameters with low error sensitivity are identified.
- Nonlinear link parameters can be identified only when triggered, that is, they can only be identified in the non-linear state.

5. Example analysis

5.1. Simulation analysis

The model of FV self-shunt excitation system is used as the model to be identified. Let the actual system parameter $K = 400$, $T_1 = 1$, $T_2 = 7$, $T_g = 3.83$, Take t as a known parameter, Identification K, T_1, T_2. The step disturbance with a unit value of 0.05 is taken as input and the output of generator terminal is taken as system output. The identification results are shown in Table 1.

Table 1. Identification result of linear link parameter of excitation system

Parameters	K	T1	T2
unit	/	s	s
Truth value	400	1	7
Identification value	398	1.05	6.97
deviation	0.5%	0.5%	0.42%

Let the limiting link $V_{AMAX}=7$, $V_{AMIN}=2$; Linear link parameters $K=400$, $T_1=1$, $T_2=7$, $T_g=3.83$. The linear link parameters are known, the step disturbance with a unitary value of 0.2 is taken as the input of the system. The VAMAX and VAMIN can be identified, and the results are shown in Table 2.

Table 2. Nonlinear link identification results of excitation system.

Parameters	VAMAX	VAMIN
unit	P.u.	P.u.
Truth value	7	2
Identification value	6.98	2.01
deviation	0.28%	0.5%

The improved DE algorithm can not only identify the linear link of excitation system effectively, but also identify the parameters of non-linear link effectively.In order to verify the superiority of the improved DE algorithm, three different artificial intelligence algorithms, namely, differential evolution algorithm, PSO and improved differential evolution algorithm, are used to identify the second-order linear system. The convergence comparison chart of the algorithm is shown in Figure 2, and the identification results are shown in Table 3.

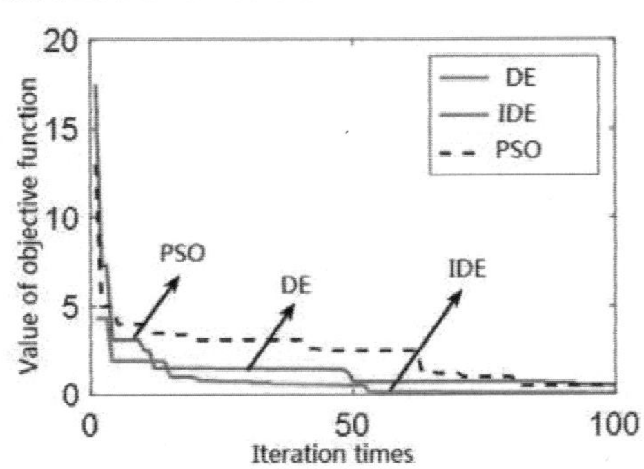

Figure 2. Objective function curves of three algorithms.

Table 3. Identification results of three algorithms.

Algorithm	Each iteration is time consuming(s)	T1 (True value is 7)	T2 (True value is 2)	Mean deviation
PSO	12.15	7.15	1.95	2.32%
DE	7.87	7.17	2.03	1.81%
IDE	8.06	6.98	2.01	0.35%

From Figure 2, it can be seen that the improved DE algorithm converges to zero in about 50 generations, and the convergence speed is obviously faster than other algorithms.In order to verify the validity of the decomposition link identification strategy, the linear and non-linear parameters of FV self-shunt static excitation system are identified step by step according to the identification process of the decomposition link identification strategy. The identification results are compared with the overall identification results, as shown in Table 4.

Table 4. Comparison of integral and link decomposition identification results.

Parameters	unit	Truth value	Decomposition Identification results	Decomposition identification bias	Global identification results	Global identificati on bias
K	/	400	380.5	0.5%	350.48	12.5%
T1	s	1	1.01	0.5%	2.13	113%
T2	s	7	6.97	0.42%	8.96	28%
Tg	s	8.38	8.37	0.1%	20.5	114%
VAMAX	p.u.	7	7.01	0.28%	8.05	15%
VAMIN	p.u.	2	1.98	0.5%	2.3	15%

5.2. Measured data identification

According to the actual data measured on site, the parameters of excitation system are identified by using improved DE algorithm and link decomposition identification strategy. The disturbance signal unit of measured data is divided into 0.05 and 0.2 step signals. The sampling frequency is 1600 Hz. The identification results of the measured data are shown in Table 5. The output of the system in case of small interference and large interference is shown in Figure 3 and Figure 4.

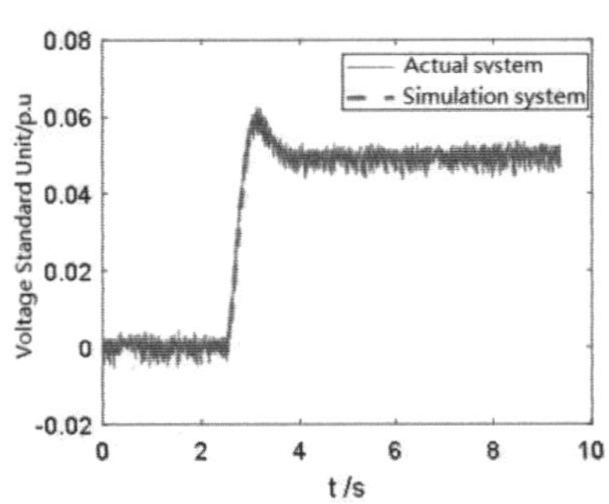

Figure 3. System output under small disturbance.

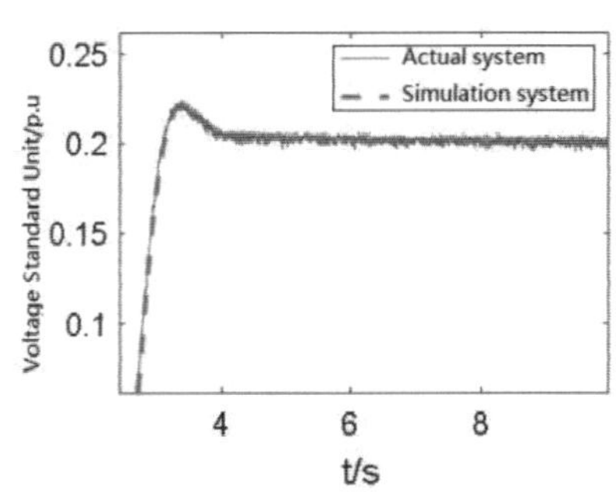

Figure 4. System output under large disturbance.

Table 5. Identification results of measured data.

Parameters	unit	Truth value	Identification results	Mean deviation
K	/	400	360.6	10%
T1	s	1	0.95	5%
T2	s	7	6.73	3%
Tg	s	8.38	8.5	1.4%
VAMAX	p.u.	7	7.21	3%
VAMIN	P.u.	2	1.9	5%

PSO, DE and IDE are used according to the decomposition strategy.The identification results and deviations are shown in Table 6. The system outputs processed by the differential evolution algorithm and the output of the actual system are compared in Figure 5 and Figure 6.

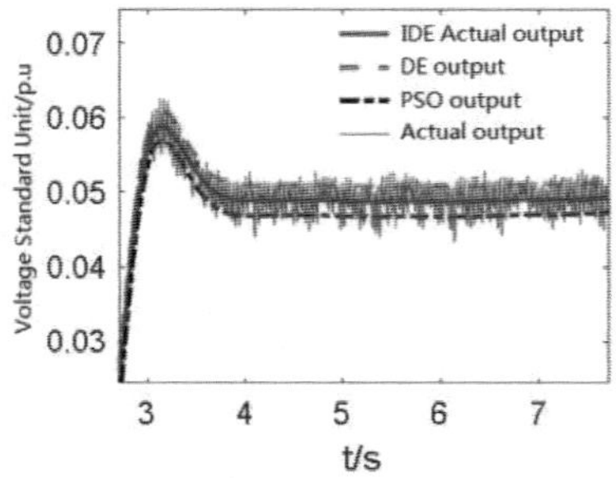

Figure 5. System output under small perturbations.

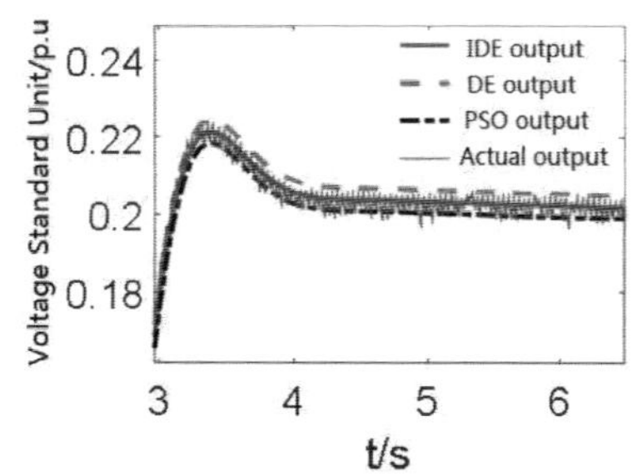

Figure 6. System output under large disturbances.

Table 6. Data identification results of different algorithms.

Parameters	Truth value	PSO	PSO bias (%)	DE	DE bias (%)	IDE	IDE bias (%)
K	400	338.5	15.5	450	12.5	360.5	10
T1	1	1.13	13	0.91	9	0.95	5
T2	7	7.43	6.1	6.1	12.9	6.73	3
Tg	8.38	6.3	14.7	8.97	7.9	8.5	1.4
VAMAX	7	6.35	9	7.81	11.5	7.21	3
VAMIN	2	1.83	9.15	2.2	10	1.9	5

From Table 6, it can be seen that the improved differential evolution algorithm and link decomposition identification strategy proposed in this paper can effectively identify the linear and non-linear link parameters of the actual excitation system of power grid.

References

[1] Ju-Sik Kim, Chang-Kyu Song, Byeong-Seok Jeon, et al. (2001) "A frequency domain identification method using total least squares" ISIE 2001. 2001 IEEE International Symposium on Industrial Electronics Proceedings (Cat. No.01TH8570) (2001): 1855-1859.

[2] Ahmed N. Abd-Alla, S. J. Cheng, J. Y. Wen, and Jing Zhang. (2006) "Model Parameter Identification of Excitation System based on a Genetic Algorithm Techniques" 2006 International Conference on Power System Technology (2006).

[3] Zhao Shuqiang, Wang Lei, Ma Yanfeng, Zhang Xigang, and Zhou Wei. (2007) "Parameter identification of nonlinear excitation system based on improved genetic algorithm" Electric Power Automation Equipment 27.7 (2007): 1-4

[4] Ahmad Darabi, Alireza Alfi, et al. (2012) "Employing Adaptive Particle Swarm Optimization Algorithm for Parameter Estimation of an Exciter Machine" Journal of Dynamic Systems, Measurement, and Control 134.1 (2012): 011013.

[5] L. M. Hajagos, and M. J. Basler. (2005) "Changes to IEEE 421.5 recommended practice for excitation system models for power system stability studies" IEEE Power Engineering Society General Meeting (2005): 1-3

[6] H. Pal Thethi, Babita Majhi, and G. Panda. " Identification of Nolinear Systems Using Inceremnetal and Diffusion Differential Evolution Algorithms" Advanced Materials Research 403-408 (2011): 3503-3509.

IOP Conf. Series: Materials Science and Engineering **752** (2020) 012018 doi:10.1088/1757-899X/752/1/012018

Simulation of Internal and External Electric Field of UHV DC Polar Bus Arrester under Pollution Conditions

Pipei ZHANG*, Wei SHI, Jie LI, Jingwen SUN and Peng WANG

State Grid Shandong Electric Power Research Institute, Jinan 250003, China

*E-mail: sddky_zpp@163.com

Abstract. Due to the adsorption effect of DC electric field on pollution particles, the pollution accumulation of DC arrester is more serious than that of AC arrester. In addition, the UHV DC polar bus arrester also has a parallel resistance pole inside, the external electric field and the internal radial electric field will be affected by the pollution distribution. In this paper, the electric field simulation model of UHV DC polar bus arrester under the condition of pollution was established, which was used to calculate and analyze the electric field on the surface of the insulating coat and the radial electric field between the resistance pole and the insulating coat when the overall pollution and the local dry belt of the arrester occur. The results showed that the overall pollution of the arrester would lead to the increase of the surface field strength and the internal radial field strength. While the value was small, it would not cause corona and flashover. When the partial dry belt occurred, it would form the partial high surface field strength and radial field strength on the surface of skirt and the air gap between the resistance piece and the insulating coat, which would cause flashover on the surface of the insulating jacket, corona discharge and penetration discharge between the resistance piece column and the insulating coat.

1. Introduction

With the continuous advancement of UHV DC transmission technology, more and more UHV converter stations are put into operation one after another, and UHV DC polar bus arrester (hereinafter referred to as 'polar bus arrester') is important overvoltage protection devices in the DC field of converter stations. The polar bus arrester is generally arranged outdoors, so the surface of the insulating jacket will be affected by local environmental pollution and weather factors such as frost, rain and snow, and a pollution layer mainly composed of salt and dust will be formed over many years.[1-3] Moreover, the adsorption of the direct current electric field on the contaminated particles makes the pollution of DC equipment more serious than that of AC equipment. Therefore, it is necessary to study and analyze the distribution of the internal and external electric fields of the polar bus arrester under the pollution condition.

It was studied in [4,5] that the pollution-resistant capacity of AC and DC arresters under different voltage levels and different pollution conditions through the form of artificial pollution test, but the test period was long, the error was large, and it is difficult to know the electric field distribution inside and outside of each section only through the monitoring of the bottom leakage current. In [6,7], the test of the contamination characteristics of the DC line insulators and the simulation of electric field under different pollution conditions were conducted. The results showed that the surface pollution distribution of the DC insulators was affected by various factors such as the position of the skirt, the

Content from this work may be used under the terms of the Creative Commons Attribution 3.0 licence. Any further distribution of this work must maintain attribution to the author(s) and the title of the work, journal citation and DOI.
Published under licence by IOP Publishing Ltd

voltage distribution and the wind speed and direction. Moreover, after the pollution layer underwent the process of water absorption, leakage current increase, heat drying and local drying zone appearing, local high field strength and flashover phenomenon would occur on the surface of the insulator. The polar bus arrester also has three columns of parallel resistors inside the insulating jacket. The change of the surface electric field of the insulating jacket and the radial electric field between the resistor and the insulating jacket caused by pollution distribution will have a certain impact on the operating condition of the arrester.

In this paper, based on the distribution characteristics of pollution in the DC field, the electric field simulation model of the UHV DC polar bus arrester was established. And the surface electric field of the insulating jacket and the radial electric field between the resistor and the insulating jacket of polar bus arrester in terms of overall pollution and local drying zone were calculated and analyzed.

2. Pollution simulation model of polar bus arrester

The $\pm$800kV UHV DC Bus Arrester is composed of five arrester sections in series, each section has a height of 2852 mm, and three resistors in parallel, which is divided into three resistor packaging units (the number of resistors column in the upper, middle and lower three units is 14, 16, 14 respectively). Resistors are pie-shaped, 106mm in diameter and 23mm in thickness. Aluminium gaskets with the same diameter and height of 20 mm are used to separate them. The schematic diagram of the overall and internal structure of the pole bus arrester is shown in figure 1. Because of the uniform potential distribution of the resistors in the polar bus arrester under DC operating voltage, the voltage distribution of each section is similar, and the influence of DC bus and other equipment is neglected, a single arrester section in the polar bus arrester is selected to establish the simulation model in this paper, which can simplify the simulation process. At the same time, the electric field simulation model of arrester surface pollution is established by adding a layer of pollution layer with a thickness of 1 mm on the outer surface of the porcelain jacket as shown in figure 2. The resistivity of the pollution layer reflects different contamination states.

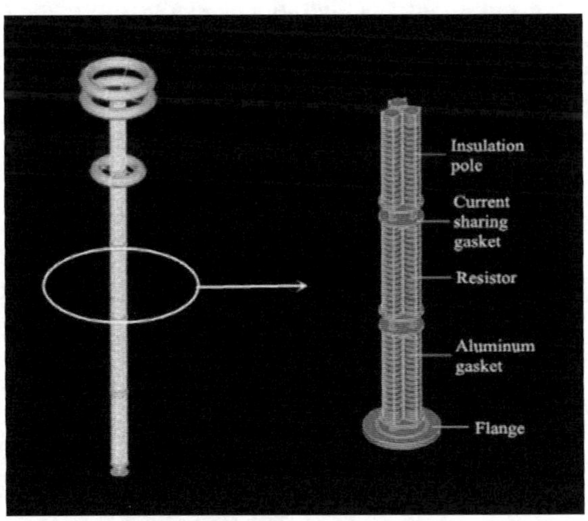

Figure 1. Structure diagram of polar bus arrester.

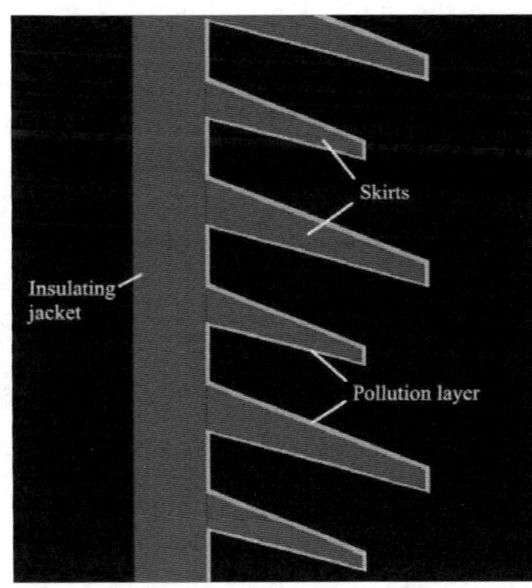

Figure 2. Pollution model of arrester.

The pollution on the jacket surface of polar bus arrester is not uniform distribution. The upper surface of skirt can be washed by rain water, whose self-cleaning effect is stronger than the lower surface, resulting in the lower surface pollution heavier than the upper surface. The salt density and ash density of the lower surface are 5-7 times as high as the upper surface.[8,9] Because DC voltage has sedimentation effect on contaminated particles, more contaminated particles will be adsorbed on the surface of insulating coat at higher field strength. The simulation results of the average field strength on the surface of each skirt of the insulating jacket of single-section arrester under the

IOP Conf. Series: Materials Science and Engineering **752** (2020) 012018 doi:10.1088/1757-899X/752/1/012018

continuous DC operating voltage of 165 kV are shown in figure 3. The field strength on the surface of the jacket is U-shaped from top to bottom, i.e., the two ends are high and the middle is low, so the pollution density on the top and bottom skirts is higher than that on the middle skirts. Pollution on the arrester jacket is mainly composed of soluble salts and insoluble materials such as silica. The greater the salt density, the stronger the water absorption, the easier the contamination layer is to be damped and the smaller the resistivity. Table 1 refers to [10-12] and the material property parameters of each part of the pollution simulation model are given in combination with the surface pollution characteristics of the DC arrester jacket.

Table 1. Material property of each component.

Component		Resistivity/(Ω • m)
Air		1014
Insulating jacket		2.5×1012
ZnO resistor		1012
Pollution on upper surface		4.2×104
Pollution on lower surface	Top skirts	35
	Middle skirts	960
	Bottom skirts	230

3. Simulation of overall pollution

According to the parameters set in table 1, the electric field distribution of the single section in polar bus arrester under 165 kV DC continuous operating voltage is simulated. The electric field distribution on the surface of the insulating jacket is shown in figure 4. The local high electric field strength appears at the apex of all umbrella skirts, and the maximum values appear at the top and bottom, which are 0.373 kV/mm and 0.296 kV/mm, respectively. This is due to the different pollution density on the upper and lower surfaces of skirts. The resistivity of the contaminated layer changes abruptly at the junction of the upper and lower surfaces. However, the resistivity of the pollution layer on the top and bottom of the insulating jacket is more different, which leads to the increase of the electric field strength at the apex of the skirt and the appearance of the maximum value.

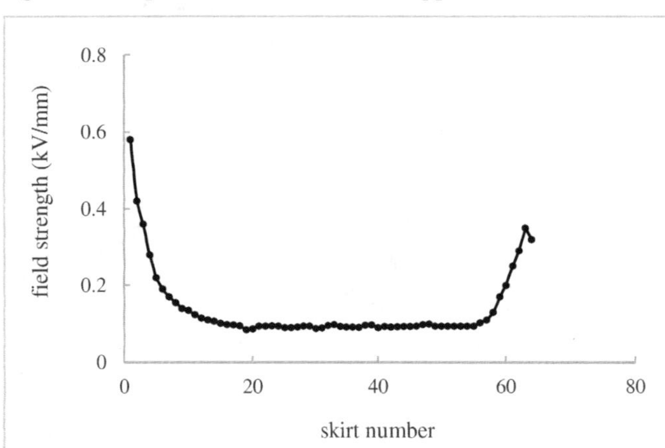

Figure 3. Electric field of the surface.

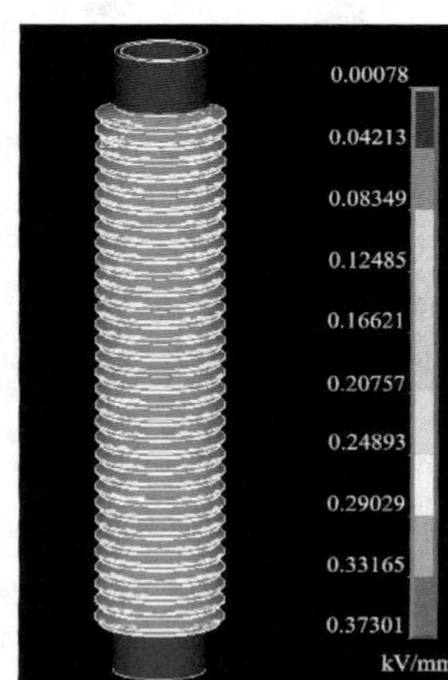

Figure 4. Surface electric field of overall pollution.

The air area between the resistor column and the insulating jacket is selected as the observation object of the radial electric field, and the whole coordinate system is adjusted to the cylindrical coordinate system. The radial field strength distribution when overall pollution and surface cleaning is shown in figure 5 and figure 6. The radial field strength of overall pollution increases with respect to surface cleaning, and concentrates at the top and bottom of the air gap, where the maximum value is 0.109 kV/mm. It can be seen from the surface potential distribution of the insulating jacket in two cases that the overall pollution causes the change of the potential distribution of the insulating jacket, which is no longer consistent with the uniform potential distribution of the internal resistor column, thus leading to the increase of the radial electric field.

No matter the maximum surface field strength or the maximum radial field strength, it does not exceed the initial corona field strength of 0.45 kV/mm on the insulator surface. Therefore, the overall pollution on the surface of the insulating jacket of the arrester will not lead to corona and flashover.

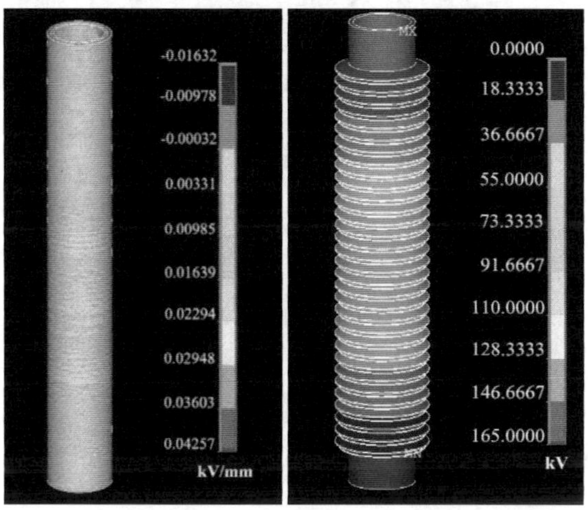

Figure 5. Radial electric field and voltage distribution of clean surface.

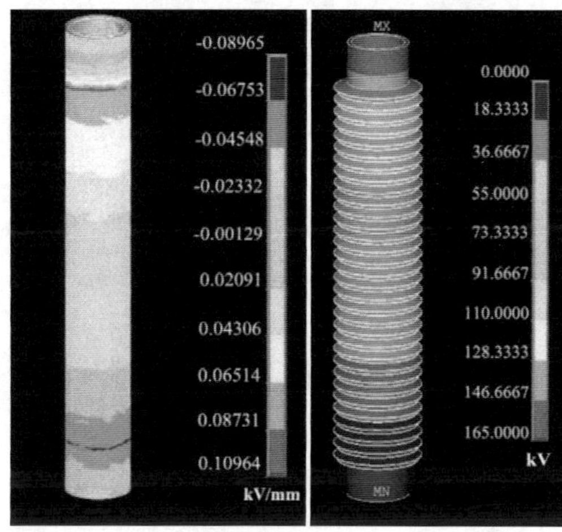

Figure 6. Radial electric field and voltage distribution of overall pollution

4. Simulation when local drying zone occurs

Polar bus arrester jacket surface pollution will lead to leakage current increasing, heating increasing, and make pollution drying to form a local drying zone. Considering that the upper surface of skirt has low pollution density, less moisture content and can accept sunlight, this paper considers that the drying zone first appears on the upper surface of skirt. The insulation performance of dry pollution is good, which can be approximately considered to be consistent with the resistivity of insulation jacket. In this paper, the internal and external electric field distribution of drying zone at the top, middle and bottom of the insulation jacket is simulated.

Figure 7 shows the distribution of field strength on the surface of insulating jacket when the local drying zone appears at different locations. It can be seen that the field strength at the drying zone is much higher than that at the rest of the jacket surface. This is because the resistivity of the local drying zone is much larger than that of the wet pollution at other parts. In the DC field, the voltage and resistivity are proportional to each other. Most of the voltage drop is borne by the drying zone, and the locally excessive field strength also occurs here. The maximum field strength of local drying zone is 8.089 kV/mm, 6.471 kV/mm and 7.990 kV/mm when it appears at the top, middle and bottom, respectively. The relationship between the magnitude and the cause is similar to that of the overall pollution in Section 3.

No matter where the drying zone appears, the maximum surface field strength produced by the drying zone far exceeds the initial corona field strength of 0.45 kV/mm on the insulator surface and the breakdown field strength of air of 3 kV/mm, which will lead to flashover discharge along the insulating jacket of the arrester and accelerate the deterioration of the insulating material.

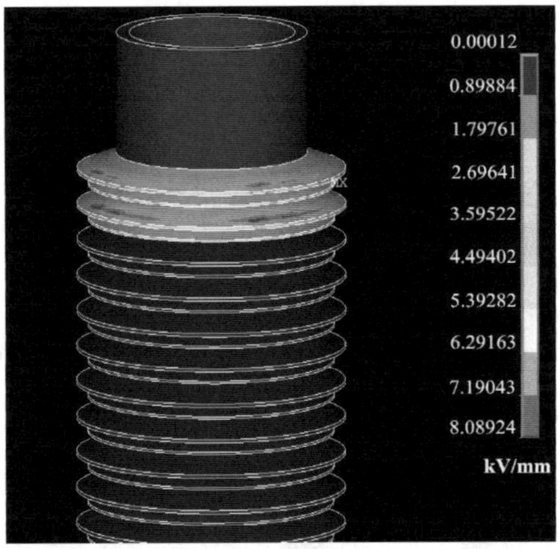

a). Drying zone at the top.

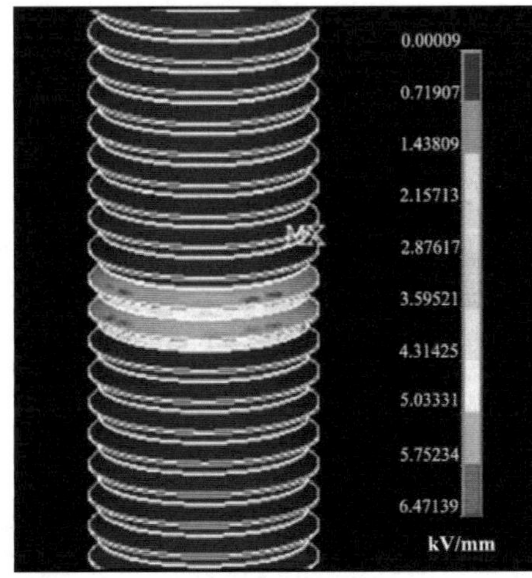

b). Drying zone in the middle.

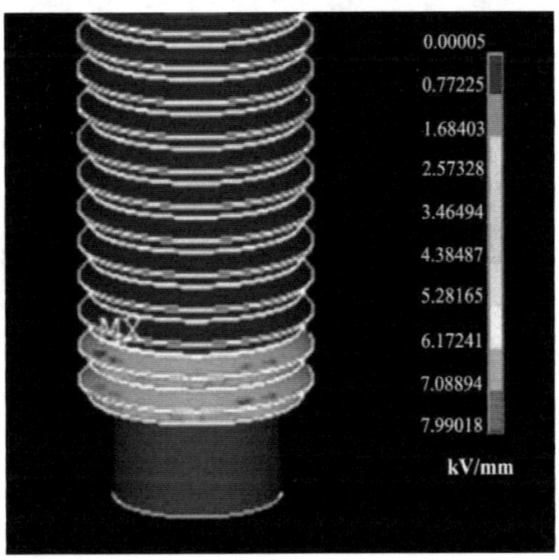

c). Drying zone at the bottom.

Figure 7. Surface electric field of local drying zone at different location.

Similarly, the air area between the resistor column and the insulating jacket is taken as the object of observation, and the radial field strength distribution is shown in figure 8 after the whole coordinate system is set as the cylindrical coordinate system. It can be seen that the larger radial electric field appears in the air region at the same location as the local drying zone, and the numerical value is shown in table 2.

The positive value indicates that the direction of the radial electric field is that the resistor points to the insulating jacket, while the negative value is opposite. The maximum radial field strength is similar when the local dry zone appears at the top and bottom, but in the opposite direction.

Further observation of the contours of the resistor column and the surface of the insulating jacket in both cases in figure 9 shows that when the drying zone appears at the top, most of the voltage is borne by the drying zone, and the contours are concentrated at the top, while the potential distribution of the resistor column remains uniform.

The potential of the resistor column at the same height is higher than that of the insulating jacket, thus forming the radial electric field from the resistor column to the insulating jacket. When the drying zone appears at the bottom, the analysis process is similar, except that the insulating jacket at the same height is at high potential, thus forming the radial electric field from the insulating jacket to the resistor column.

IOP Conf. Series: Materials Science and Engineering **752** (2020) 012018 doi:10.1088/1757-899X/752/1/012018

When the drying zone appears in the middle, as shown in figure 9, the contours of the outer surface concentrates in the middle, and the potential of the upper resistor column is higher than that of the insulating jacket of the same height, while the lower part is opposite. Therefore, two radial field strengths with opposite directions and similar values are formed on the upper and lower sides of the drying zone. It can be seen that the inconsistency of the potential distribution between the resistor column and the insulating jacket is the main reason for the increase of the radial field strength when the local drying zone appears on the surface of the skirt.

When the local drying zone appears in the top and the bottom, the maximum radial field strength exceeds the breakdown field strength of air, which will lead to the breakdown of the air gap between the resistor column and the insulating jacket, and endangering the safe and stable operation of the arrester.

When the local drying zone appears in the middle, the air gap at both the upper and lower sides share the inconsistency of the internal and external potential distribution, so the maximum radial field strength formed is smaller than that of drying zone appearing at the top and bottom, but it also exceeds the initial field strength of corona on the surface of insulating materials. Long-term corona discharge also accelerates the aging of resistors and insulating materials.

Table 2. Maximum radial field strength of local drying zone at different location.

Location of drying zone	maximum radial field strength (kV/mm)
Top	7.892
Middle	2.495/-2.612
Bottom	-7.153

5. Conclusion

In this paper, the single section of arrester is taken as the research object, and the electric field simulation model of UHV DC polar bus arrester under the condition of pollution is established. The electric field on the surface of skirt and the radial electric field between resistor and insulating jacket are calculated and analyzed when overall pollution and local dry zone occurs. The following conclusions are drawn:

1) When polar bus arrester is overall polluted, the surface and internal radial field strengths of the insulating jacket increase due to the different pollution density at different locations on the surface of the insulating jacket, and there are maximum values at the top and bottom, but the smaller values do not cause corona and flashover;
2) When the local drying zone appears on the surface of the insulating jacket, the surface electric field strength at the drying zone increases sharply, far exceeding the breakdown electric field strength of the air, which will lead to flashover discharge along the surface of the insulating jacket of the polar bus arrester and accelerate the deterioration of the insulating material;
3) The larger radial electric field appears at the same height as the local drying zone. When the drying zone appears at the top and bottom, the maximum radial electric field strength is similar and opposite, but its value exceeds the breakdown electric field strength of air. When the drying zone appears in the middle, two opposite and similar radial electric field strengths are formed at the upper and lower sides of the drying zone, and its value is smaller than breakdown electric field strength of air, but exceeds the initial corona field strength on the surface of insulating materials. It can be seen that the flashover of insulating jacket surface, corona discharge and penetrating discharge between resistor column and insulating jacket will occur on the surface of polar bus arrester after being polluted, wetted, heated and dried. The deterioration of resistor and insulating material will be an important factor to endanger the safe and stable operation of arrester.

IOP Conf. Series: Materials Science and Engineering **752** (2020) 012018 doi:10.1088/1757-899X/752/1/012018

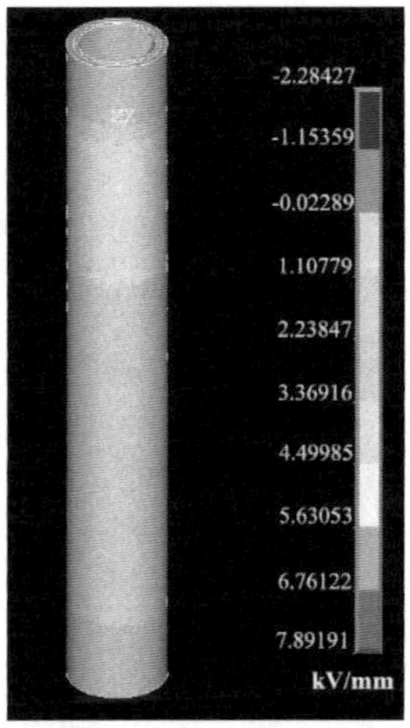

a). Drying zone at the top.

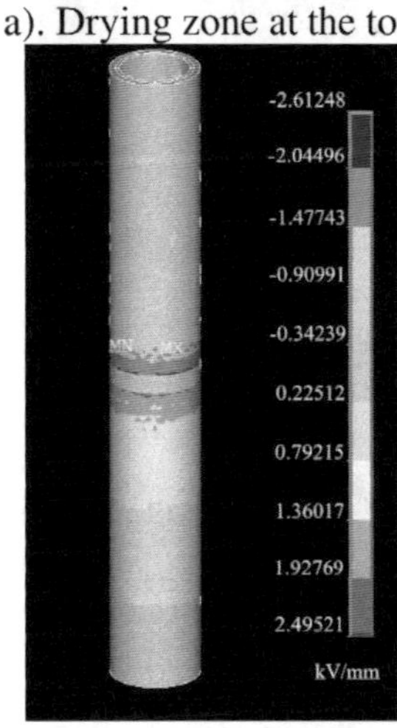

b). Drying zone in the middle.

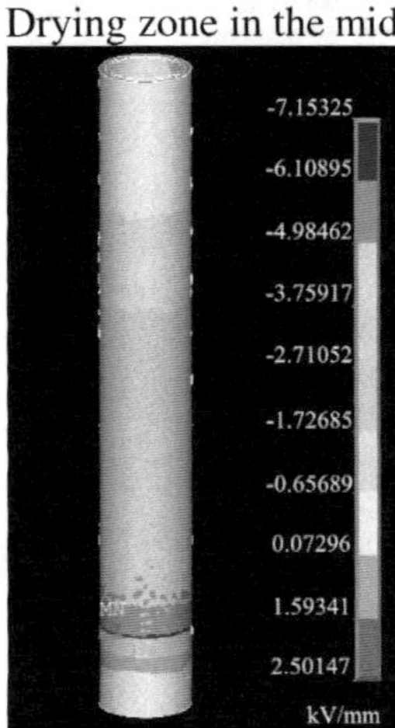

c). Drying zone at the bottom.

Figure 8. Radial electric field of local drying zone in different location

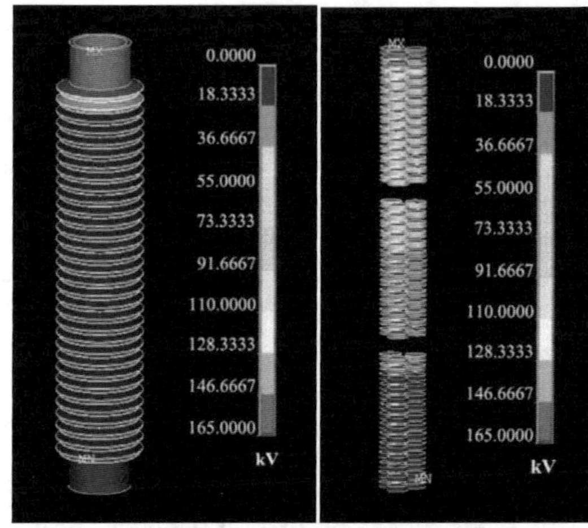

a). Drying zone at the top.

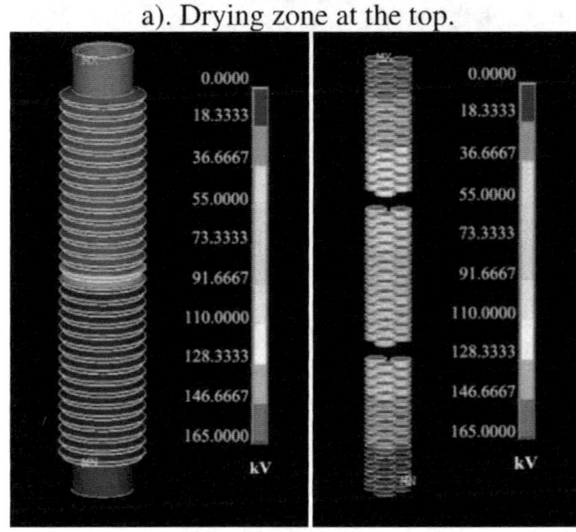

b). Drying zone in the middle.

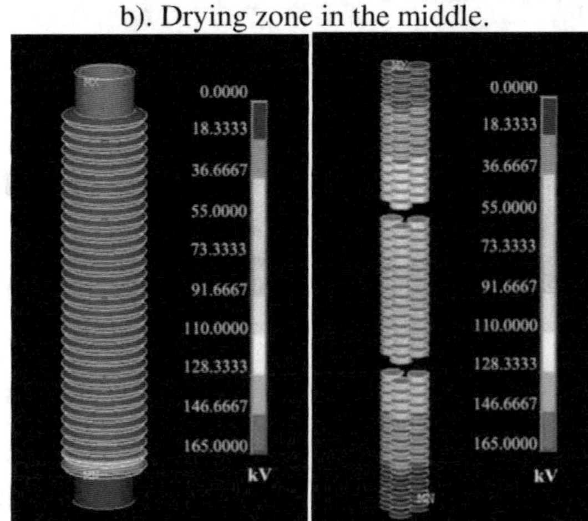

c). Drying zone at the bottom.

Figure 9. Voltage distribution of local drying zone in different location

IOP Conf. Series: Materials Science and Engineering **752** (2020) 012018 doi:10.1088/1757-899X/752/1/012018

References

[1] ZHOU P H, ZHAO J and LV J Z 2009 Arrester Protection Scheme and Arrester Parameters Selection for ±800kV YunGuang UHV DC Converter Station *High Voltage Engineering* **35(11)** 2603-11
[2] ZHANG X X, LI Z and ZHAO D C 2010 Key Technologies Applied in Research and Development of HVDC Surge Arresters *Power System Technology* **34(8)** 150-154
[3] HU S H 2016 Research on Pole Bus Metal Oxide Surge Arrester for ±800kV UHVDC Converter Station *High Voltage Apparatus* **34(8)** 34-45
[4] ZHANG B Y, SU N and LV X B 2014 Research on the Pollution-Resistant Capacity of AC/DC MOA *Insulators and Surge Arresters* **1** 79-84
[5] METWALLY I A, GASTLI A and AHED M M 2006 Surge Arresters Performance under AC Artificial Pollution and Impulse Conditions: a Comparative Investigation *Electric Power Components and Systems* **34(11)** 1243-1251
[6] LU W H, CHEN B and LIN S Y 2018 Anti-interference Research on Live Test of DC Arresters in Converter Station *High Voltage Apparatus* **54(9)** 191-196
[7] LIU Z X, LAN L and CHEN Y 2017 Partial Discharge Test and Electric Field Simulation of Polluted Composite Material Tower *Insulation Materials* **50(1)** 61-67
[8] YUAN J H, JIANG X L and SHU L C 2007 Influence of Salt/Non-soluble Deposit Density on AC Artificial Pollution Flashover Performances of Various Types Insulators *Proceedings of the CSEE* 27**(6)** 96-100
[9] WANG B, LIANG X D and ZHANG Y B 2009 Natural Pollution Test of Composite and Porcelain Insulators Under AC and DC Stress *High Voltage Engineering* **35(9)** 2322-2328
[10] ZHANG B Y, LI G F and ZHANG C X 2011 Research on the Internal Temperature rise of MOA under Pollution Condition *High Voltage Engineering* **37(8)** 2065-2072
[11] AN C C, LIU H and WANG Q 2009 Calculation of Electric Potential and Field Distribution of UHVDC Bus Bar MOA *High Voltage Apparatus* **45(5)** 9-15
[12] TIAN J H, SHA Y C, ZHOU Y X 2011 Simulation of AC and DC Electric Field for 800kV Grade High Voltage Disconnector *High Voltage Engineering* **37(5)** 1216-23

IOP Conf. Series: Materials Science and Engineering **752** (2020) 012019 doi:10.1088/1757-899X/752/1/012019

Design of Piston Gauge Verification Data Processing System Based on LabVIEW

Yao Yao Cui[1,a], Bo Sun[2,b], Hong Guang Liu[1,c] and Zhe Wang[1,d]

[1]Tianjin Institute of Metrological Supervision and Testing, Tianjin 300192, China
[2]Tianjin Bicycle Research Academy, Tianjin 300111, China

Email: [a]13820325752@163.com; [b]165288912@qq.com; [c]18622591898@163.com; [d]lhgredlight@sina.com

Abstract: In order to check the weight of the quality of piston and its connecting parts more conveniently, this paper designs a set of data processing system for piston gauge measurement and verification based on LabVIEW. The system uses LabVIEW development platform to establish a visual human-machine interface. According to the requirements of JJG59-2007 National Verification Regulation of Piston Pressure Gauge, a set of data processing system is designed and completed. The system can accurately and effectively calculate the special weights of Special Weight, Piston and Connecting Parts of Piston Gauge according to the verification data. It greatly improves the work efficiency. The research lays a foundation for adding piston position indication device and automatic generating and exporting of original records in the system.

1. Introduction

Piston pressure gauge is a measuring standard that uses the principle of hydrostatic balance (that is, the fluid pressure acting on the effective area of the piston is balanced with the gravity of its load). It is a high accuracy, high reproducibility and highly credible standard pressure measuring instruments[1-3]. According to the requirements of the national verification regulations of JJG59-2007 "Piston Pressure Gauge", after the instrument completes the effective area verification of the piston, there will be a large amount of raw data to calculate the effective area of the piston, the special weight, the quality of the piston and its joints according to the formula. . Due to the complex formula and large amount of data, the calculation is easy to make mistakes and the work efficiency is low. For this reason, the intelligent data processing system can greatly improve the accuracy, reliability and efficiency of the verification results[4-5].

In recent years, with the development of artificial intelligence, virtual instruments have been more widely used. Instead of the control panel of a conventional instrument, the virtual instrument uses a computer's powerful graphical environment to create a graphical soft-faced panel on a computer screen instead of a conventional traditional instrument panel using a visual graphical programming language and platform. The soft panel has knobs, switches, indicators and other control components similar to the actual instrument. During operation, the user operates the soft panel with a mouse or keyboard to verify the communication and operation of the instrument.The development language based on LabVIEW graphics is simple and easy to understand, and it is convenient to adjust and

Content from this work may be used under the terms of the Creative Commons Attribution 3.0 licence. Any further distribution of this work must maintain attribution to the author(s) and the title of the work, journal citation and DOI.
Published under licence by IOP Publishing Ltd

maintain subsequent software. Therefore, this paper uses LabVIEW virtual instrument development platform to establish a simple and friendly visual human-computer interaction interface, and designs an intelligent piston pressure gauge verification data processing system. The system uses LabVIEW visual programming technology to establish a human-computer interaction interface. Through the raw data processing and analysis obtained by the verification, the piston effective area, special weight, piston and its connector quality are obtained [6-8].

2. Data processing system design

The system uses the LabVIEW development platform developed by NI National Instruments to realize the functions of data display, analysis and processing, which is easy to operate and maintain. The overall design of the system is as follows: the piston and the standard piston pressure gauge to be tested are installed on the same calibrator (or the standard piston pressure gauge and the piston pressure gauge to be tested are connected through the pipeline), and the piston is adjusted. Vertical position. According to the hydrostatic balance method, the area of the piston gauge to be inspected is compared with a standard pressure gauge. Since the standard piston type pressure gauge and the piston system of the tested piston type pressure gauge have basically the same material and shape, the laboratory adopts the initial balance method. After the verification according to the regulations, the detected piston is calculated from the obtained raw data. The effective standard area of the pressure gauge piston, the experimental standard deviation of the effective area value, the piston effective area limit error, the special weight, the piston and the joint quality[9-12]. The overall design of the system is shown in Figure 1:

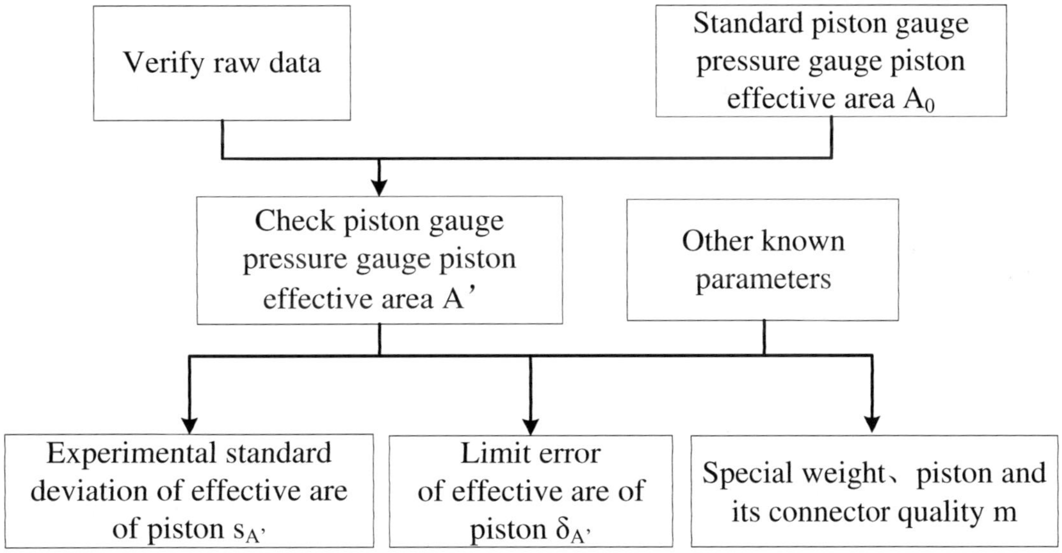

Figure 1. Overall system design

In order to realize the function of data processing, the block diagram is divided into two parts. Firstly, the effective area and effective area value of the piston of the piston type of the tested piston are calculated by the obtained raw data and the effective area of the piston of the known standard piston gauge. Experimental standard deviation, piston effective area limit error; secondly, through the piston piston gauge effective area and other known parameters, such as measured pressure value, ambient air density, special weight, piston and connector material density and piston The gauge weight, piston and connector quality are calculated by the gravitational acceleration of the pressure gauge.

The software running environment is LabVIEW 8.5 under Windows XP, which consists of a front panel and a block diagram. Using the system's own display controls, input controls, execution of

IOP Conf. Series: Materials Science and Engineering **752** (2020) 012019 doi:10.1088/1757-899X/752/1/012019

process control loops, etc. to achieve the experimental standard deviation of the effective area value and the calculation of the special weight, piston and connector quality.The rear panel block diagram is shown in Figure 2 and Figure 3 below:

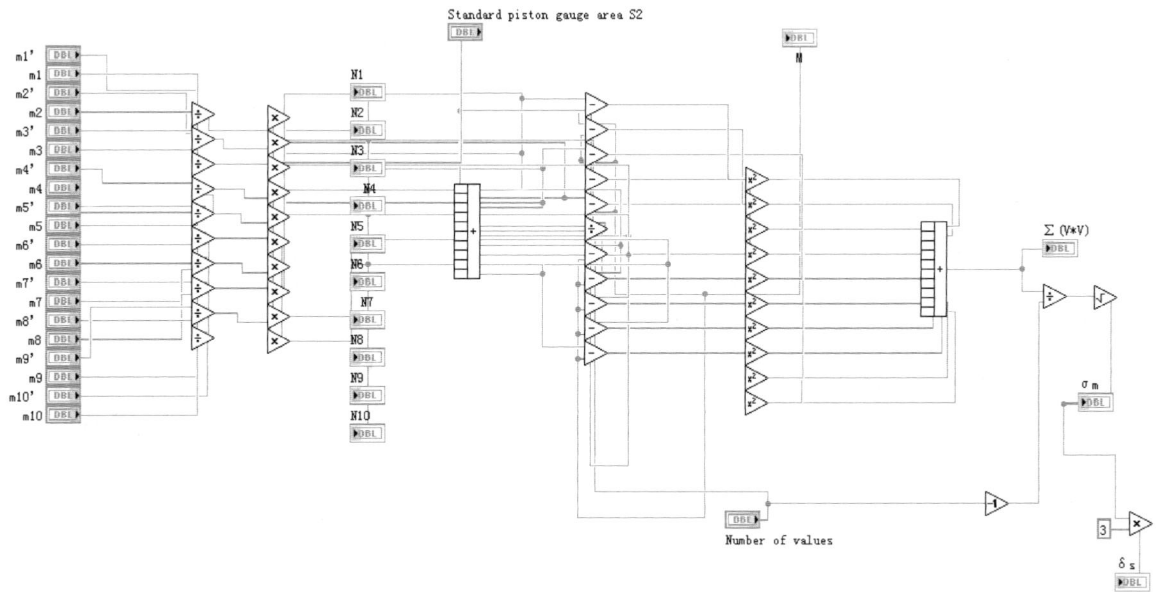

Figure 2. Block diagram of effective area of piston piston gauge

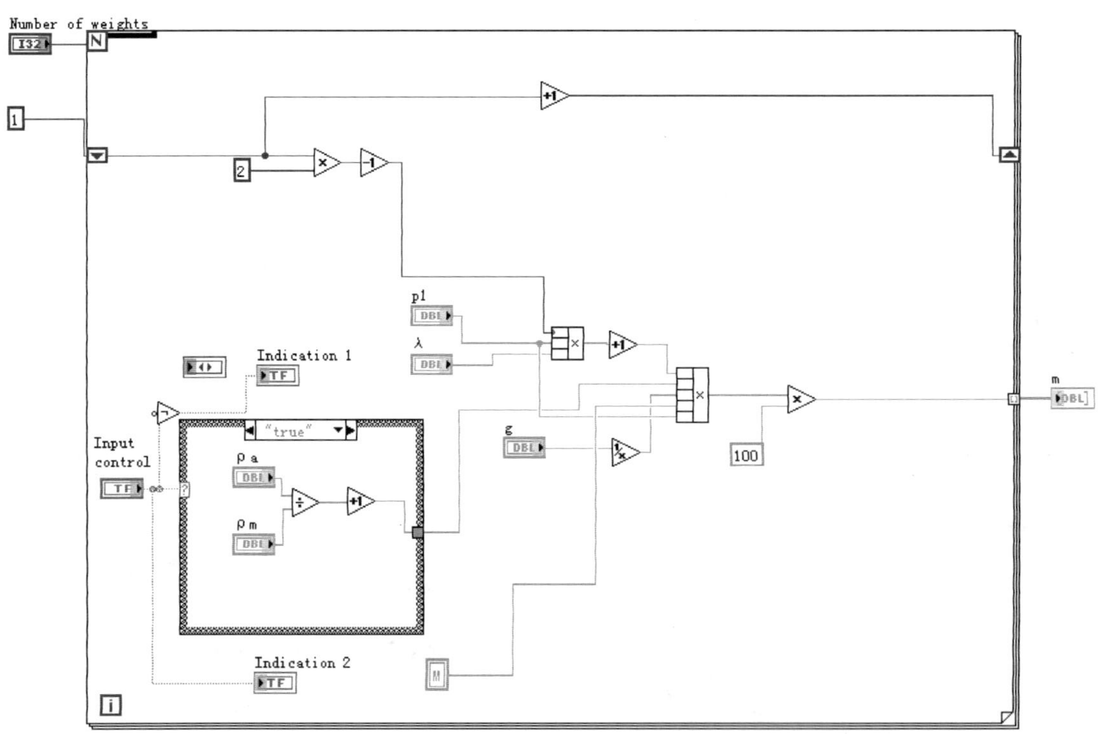

Figure 3. Special weight, piston and connector quality block diagram

3. Experiment and analysis

The system has a friendly human-machine interface, realizes raw data input, and displays the data results after analysis and processing in real time, which is convenient for operation and maintenance. The raw data and the calculated data are input and displayed by the control panel to call the program calculation, and the experimental standard deviation of the effective area value and the calculation of the special weight, the piston and the connector quality are realized, and the input and output changes are visually displayed. The control interface is shown in Figure 4 below:

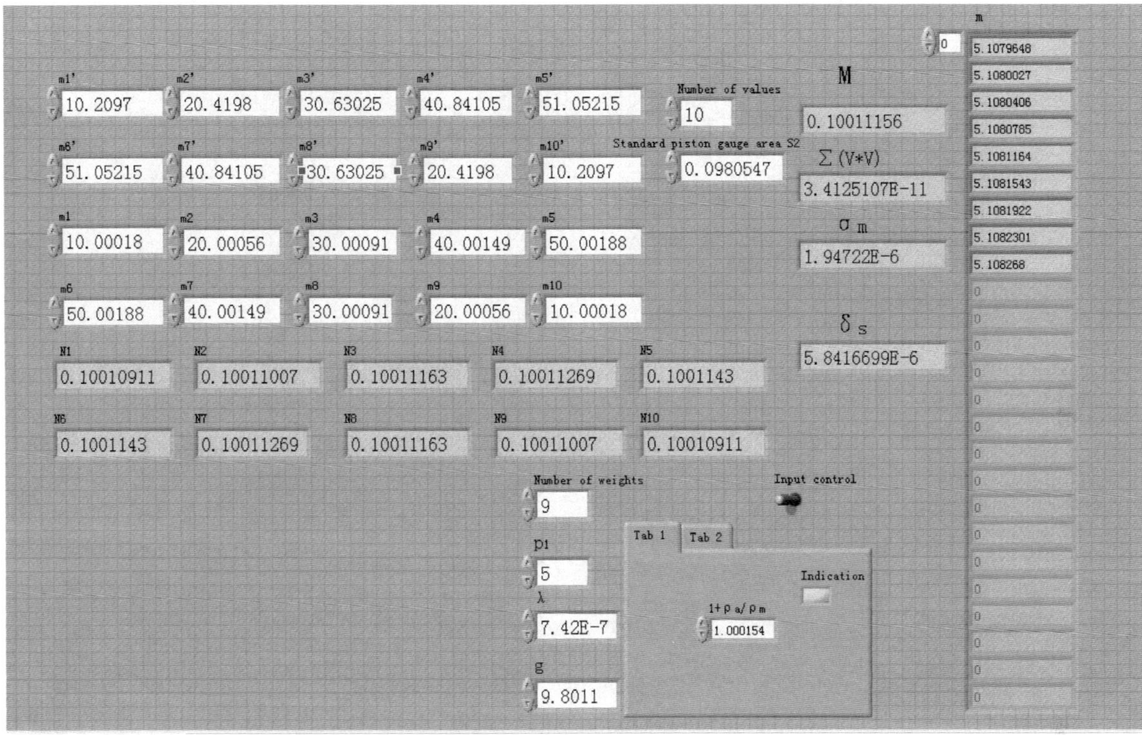

Figure 4. Data processing front panel interface

The verification raw data input through the control panel, the effective area of the standard piston gauge piston, the number of weights, the measured pressure value, the pressure deformation coefficient of the piston-cylinder assembly, the gravity acceleration at the place of use, the ambient air density, the weight material Density, etc., the running program realizes the calculation of the effective area of the piston of the tested piston gauge, the experimental standard deviation of the effective area value, the piston effective area limit error, the special weight, the piston and the joint quality, wherein the white box input is known. The data, gray box visually shows the results.During the inspection of the piston gauge, the raw data obtained can be entered into the data processing front panel during the inspection. After the verification is completed, click the Run button, and the verification result will be displayed directly on the front panel interface.

In the experiment, the calibration data of a piston pressure gauge with an accuracy level of 0.02 and a measuring range of (1~60) MPa was processed. The comparison between the effective area of the tested piston pressure gauge and the 5MPa special weight is shown in Table 1 below. The comparison results are completely correct, and the visualization operation is simple and the work efficiency is greatly improved.The traditional method takes about 30 minutes to complete a calculation on the premise of rich verification experience, and this calculation method only takes 1 second after completing the input of raw data, which greatly improves the work efficiency and reduces the cost Labor intensity.

Table 1. Comparison of calculation results

Project		LabVIEW calculation	Traditional calculation method
Checked piston pressure gauge effective area (cm^2)		0.10011156	0.10011165
5MPa special weight(Kg)	1st piece	5.10796	5.10797
	2nd piece	5.10800	5.10801
	3rd piece	5.10804	5.10805
	4th piece	5.10808	5.10808
	5th piece	5.10812	5.10812
	6th piece	5.10815	5.10816
	7th piece	5.10819	5.10820
	8th piece	5.10823	5.10824
	9th piece	5.10827	5.10827
	10th piece	5.10831	5.10831

4. Conclusion

In this paper, the data processing system of piston pressure gauge measurement and verification is studied. The LabVIEW development platform is used to establish a visual human-machine interface, and a data processing system is designed and completed. Experiments show that the system can accurately and effectively calculate the effective area of the piston of the piston gauge under test, the special weight, the piston and its joints according to the verification data, which greatly improves the working efficiency. In the next step, you can consider adding multi-channel data acquisition and signal processing to the system hardware design. Adding modules such as ambient temperature detection, piston rotation speed detection, and piston position indication device to the system can complete the verification environment in the piston pressure verification project. The work of the piston rotation duration test, the piston drop speed test and other projects basically cover all the verification items. In addition, the original record automatic generation and export function is added in the system software design, the verification data can be retained, the original issuance and the verification certificate are issued, which will greatly improve the work efficiency and the accuracy of the measurement, and ensure the accuracy and stability of the value transmission.

References

[1] LI Hai-bing.Discussion on Some Problems in the Process of Piston Pressure Gauge Verification[J].Measurement Technique, 2018(01):46-48.
[2] FAN Wei-meng, ZHANG Shen, HAN Fei, WU Hong-wei.Verify the Effective Area of the Piston by Direct Balance Method[J].China Metrology, 2014(09):105-107.
[3] LI Hai-bing, YUE Jin, ZHUO Hua, CHEN Wu-qin.Discussion on the Special Weight of Piston Pressure Gauge[J].Metrology & Measurement Technology, 2013(04): 77-80.
[4] LI Yan-hua.Problems in the Calculation and Correction of the Weight of the Piston Gauge[J].Measurement Technique, 2007(04): 50-52.
[5] HAN Fei, GUAN Wei-jun, FAN Wei-meng.Analysis of the Mass Calculation of the Piston Weight Gauge[J].Metrology & Measurement Technique, 2014(05):14-16.

[6] WANF Xiao-qi.The Application of a Computer to the Calibration of a Deadweight Pressure Tester[J].Industrial Instrumentation & Automation,2000(02):48-49.
[7] XIE YAN.Design and Implementation of Rail Transit Data Acquisition System Based on LabVIEW[J].Electronic Design Engineering,2019(15):82-85.
[8] ZHAO YAN.Design and Implementation of USB Data Acquisition System Based on LabVIEW[D].BeiJing: Beijing University of Chemical Technology, 2014.
[9] WANG Li,SHENG Xiao-yan.Piston-Cylinder Assembly Design of the Digital Piston Pressure Gauge [J].Measurement & Control Technology,2012(07):13-15.
[10] ZENG Lin,ZHOU Si-qing,LONG Yan-tao.Piston Pressure Gauge Piston Based on Equal Mass Increment Method.Effective Area Calibration Methodology[J].Measurement Technique, 2019(08):14-17.
[11] YUE Jin,YANG Yuan-chao.Comparative Analysis of Direct Balance Method and Initial Balance Method in Effective Area Determination of Piston[J].Measurement Technique, 2010(12):59-61.
[12] YAO Li-hui,HE Gong-jian,ZHANG,Sheng-yu.Research of the Manage Method in the Digital Press Calibration[J].Metrology & Measurement Technique,2010(11):21-25.

IOP Conf. Series: Materials Science and Engineering **752** (2020) 012020 doi:10.1088/1757-899X/752/1/012020

Research on Calibration Method of Temperature Field for Large Sealed High Temperature Equipment

Wang Zhe[1, *], Cui Yao-yao[1], Yu Song-lin[1], Fan Ying[1], Sun Hao[1] and Wang Xiao-dan[1]

[1]Tianjin Institute of Metrological Supervision and Testing, TianJin, CHN

[*]18622591898@163.com

Abstract. Large sealed high temperature equipment usually refers to heat treatment equipment with a volume of more than $0.15m^3$, sealed structure and internal heating and constant temperature function. It is widely used in the fields of biochemistry, chemical pharmacy, medical and health, agricultural research, automobile and military industry. It`s a very important processing equipment. The accuracy of temperature field and vacuum sealing directly affect the quality of products. Taking the hot-pressing tank as an example, the furnace temperature tracker and wireless pressure module are used to calibrate it on the spot. According to the calibration results of temperature field distribution and temperature field performance, whether it meets the process requirements is analyzed.

1. Introduction

The large-scale sealed high-temperature equipment generally refers to thermal treatment equipment with volume over $0.15m^3$ [1], sealed structure, internal heating and constant temperature functions. Some devices are provided with vacuum pressure pump to create pressure or vacuum operating environment. It is an extremely important processing equipment which has been widely used in research fields such as biological chemistry, chemical and pharmaceutical engineering, health care, agricultural research, automobile and military industry. The product quality directly depends on accuracy of thermal field and vacuum tightness. [2] However, nowadays, there is not uniform measurement method for thermal field in China. Therefore, based on related standards [3], furnace temperature tracker and wireless pressure module are used to calibrate uniformity of thermal field, temperature deviation and fluctuation, vacuum and pressure tightness in working area, providing technical bases for industrial production and scientific research. [4]

2. Equipment under test

We take military autoclave for an example. Autoclave is a set of system equipment [5] that can realize timing sequence and real-time online control for process parameters such as temperature, pressure, vacuum and cooling, therefore, being an important processing equipment in many industries. See Figure 1 for its structure. It is mainly constituted by tank, sealing device, heating system, pressure system, automatic control system and safety mechanism. The tank is designed with length of 7m and height of 3.5m, of which internal working area is similar to shape of housing, being a 6m×Φ3m horizontal cylinder. Such working area is set with constant temperature range from ambient temperature to 400°C, general working pressure of 0.3MPa and top working pressure of 1.5MPa. Besides, the temperature uniformity is ±2°C and deviation of control pressure does not exceed 10kPa.

Content from this work may be used under the terms of the Creative Commons Attribution 3.0 licence. Any further distribution of this work must maintain attribution to the author(s) and the title of the work, journal citation and DOI.
Published under licence by IOP Publishing Ltd

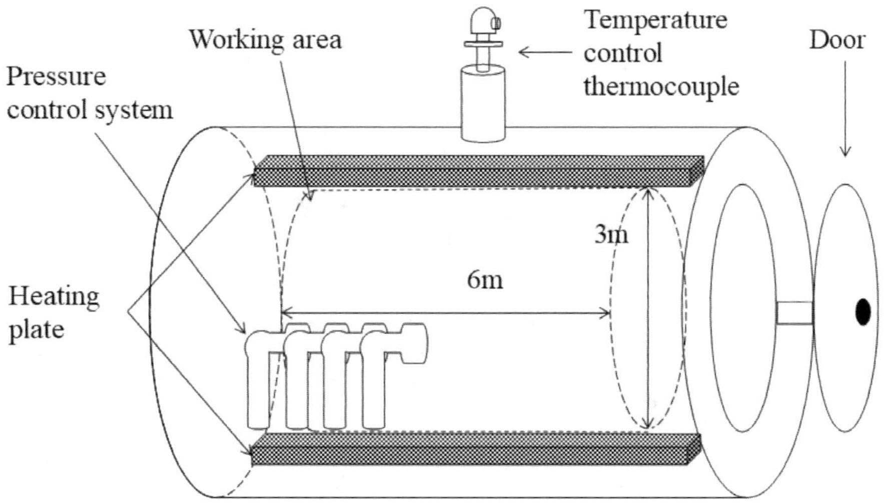

Figure 1. Structural diagram of autoclave

3. Measurement devices

3.1. Furnace temperature tracker

The measurement device is constituted by furnace temperature tracker, insulated box, phase change module, sheathed thermocouple and wireless transmission module. See Table 1 for technical indexes.

Table 1. Technical indexes for components of temperature measurement devices [6]

Name of components	Model	Technical indexes
Furnace temperature tracker	PTM1020	Size: 200mm×98mm×20mm; 20 channels; measurement accuracy: ±0.3°C; sampling interval: Min. 0.5s
Thermal box	Phoenix TS02	Size: 550mm×400mm×300mm; Upper operating temperature: 1100°C
Phase change module	Phoenix PCA	Size: 280mm×145mm×75mm
Sheathed thermocouple	K	Length: 5m; outer diameter: 1mm; range of measurement: -40°C~1100°C; grade of accuracy: Grade 1
Wireless transmission module	TXR-1000	Transmission frequency: 2.4Ghz; maximum transmission distance: 150m

The insulated box is designed with high-temperature resistant and stainless steel housing and filled with metallic oxide, fiber and high-temperature fabrics, porous and thermal insulation materials with thickness of 150mm. The internal groove is placed with metallic phase change module, while the box is fastened by two parts. When the buckle is tightened, sheathed thermocouple can extend out of the box from gaps on both sides. Due to heat absorption function, materials in phase change module will change from solid-state phase to liquid state and absorb thermal energy [7] under about 58°C. Besides, temperature will remain for a certain period to extend measurement time for furnace temperature tracker under high-temperature environment as much as possible. See Figure 2 for picture of thermal box and phase change module. The maximum operating temperature of monomer of furnace temperature tracker is 80°C [8]. See Figure 3:

IOP Conf. Series: Materials Science and Engineering **752** (2020) 012020 doi:10.1088/1757-899X/752/1/012020

Figure 2. Picture of thermal box and phase change module

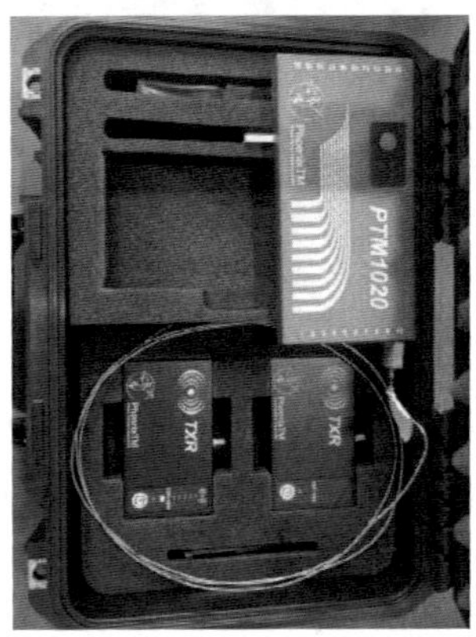

Figure 3. Picture of furnace temperature tracker

In addition to above hardware equipment, supporting testing software is also provided, in which corrected temperature values of calibrated sheathed thermocouple can be entered, and 20 channels can be calibrated one by one as per each temperature point.

3.2. Wireless pressure module

TMI wireless pressure module is used to measure pressure parameters. See Figure 4. For long-term operation, the module is measured with temperature range of (0~140)°C, measurement range of 3kPa~1.5Mpa, allowable error of ±1.2kPa, and over 40,000 stored data. [9] Besides, its tolerance does not exceed one fifth of the equipment under test and measurement accuracy satisfies with requirements. The data are recorded under sealed working environment and will be read via software system after test.

Figure 4. Picture of wireless pressure module

4. Methods of measurement

4.1. Measurement positions

Autoclave is provided with three internal separators on which uniform holes are designed. It is generally used for storing workpieces and can be used as temperature measurement rack when its temperature field is measured. Based on size and shape of internal working area in autoclave and temperature measurement position, in principle, temperature measurement positions shall be uniformly distributed in working area of autoclave, while measurement terminal of thermocouple does not contact with autoclave body or heating elements, including two important positions, 1. Geometric center of working area; 2. Measurement terminal of temperature control thermocouple in autoclave.

See Figure 5 for positions of insulated box and wireless pressure module. The wireless pressure module is positioned in the geometric center, while insulated box is positioned in the interlayer for distribution of sheathed thermocouple. Based on furnace mouth, measurement positions include "front", "middle" and "rear" face and "front" and "rear" distributions are same. See Figure 6 for distribution position of each face and distance between two points. Position 7 refers to position of measurement terminal for temperature control thermocouple and Position 11 refers to geometric center of working area for furnace body. [10]

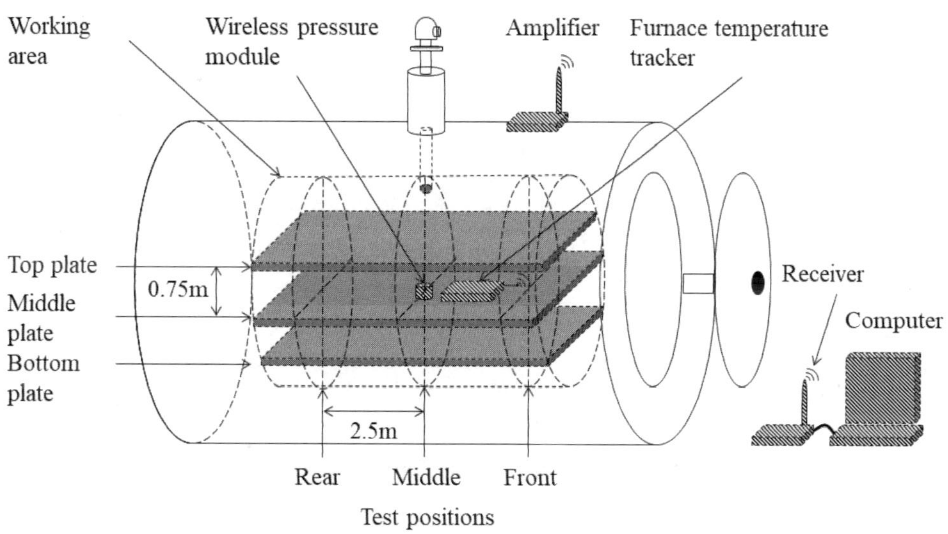

Figure 5. Diagram of measurement positions

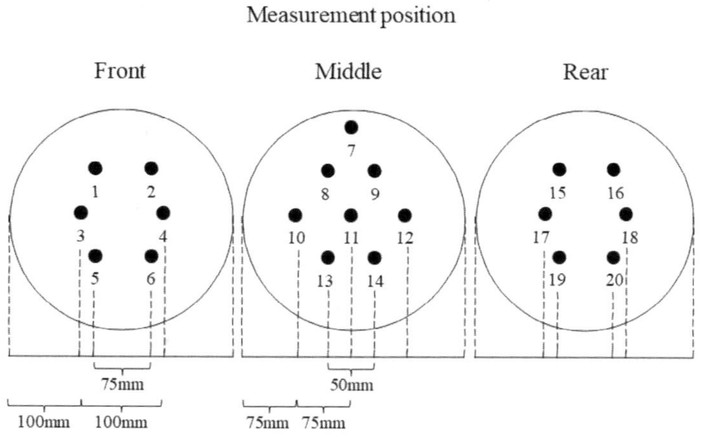

Figure 6. Distribution diagram of measurement positions

IOP Conf. Series: Materials Science and Engineering **752** (2020) 012020 doi:10.1088/1757-899X/752/1/012020

4.2. Measurement procedure

The distribution of thermal field of autoclave is measured under unloaded condition. The measurement temperature is set to upper limit of 400°C. The upper limit of heat resistance for wireless pressure module is 140°C, which is far lower than measured temperature of thermal field. Therefore, automatic control system is closed and changed to manual mode. Besides, measurement for pressure indexes shall be executed firstly and then temperature shall be controlled independently to conduct measurement for thermal field.

According to operation of equipment, it will take 2h for autoclave to achieve thermal balance from ambient temperature to 400°C and will take 3h from 400°C to ambient temperature. Autoclave is provided with internal cross-ventilation equipment. The temperature measurement equipment shall be placed under high-temperature environment for measurement, so that measurement time shall be strictly controlled in order to prevent damage of furnace temperature tracker under high temperature. According to specification, for temperature measurement equipment constituted by various components, upper limit of temperature for normal operation is 1100°C and duration does not exceed 30min. The high-temperature resistant time for measurement equipment is achieved through weighted calculation for five conditions as below: 1.Maximum temperature; 2. Time of duration; 3. Average temperature; 4. Working environment (e.g., carburizing atmosphere or static air); 5. Other factors such as handling of devices. When phase change module operates with low speed in convection air environment with maximum temperature of 400°C and average temperature of about 240°C, normal operating time does not exceed 6h, so that temperature measurement time shall be strictly controlled within 1h.

We take wireless pressure module out after pressure measurement; put measurement devices in autoclave and complete wiring for sheathed thermocouple as required; start online and real-time data acquisition after heating up and maintaining constant temperature for 2h and set acquisition interval to 3min/times. After data acquisition, temperature is declined, and measurement equipment is taken out when requirements are complied with.

5. Measurement results

After excluding abnormal values [11], data collected for 1h are divided into 20 groups with 20 values in average for each. See Table 2 for distribution of thermal field of autoclave after calculation of average value.

Table 2. Distribution results for thermal field of autoclave

Setting temperature /°C	Display temperature /°C	Actual average value/°C				
400.0	400.0	Position 1	Position 2	Position 3	Position 4	Position 5
		401.0	400.9	400.4	400.7	400.7
		Position 6	Position 7	Position 8	Position 9	Position 10
		400.7	400.6	400.8	400.9	400.7
		Position 11	Position 12	Position 13	Position 14	Position 15
		400.3	400.5	400.8	400.8	400.7
		Position 16	Position 17	Position 18	Position 19	Position 20
		400.8	400.6	400.4	400.4	400.5

For above data, according to formula (1) & (2), uniformity of furnace temperature "$\Delta\theta_+$" & "$\Delta\theta_-$" can be calculated; according to formula (3) & (4), stability of furnace temperature "δ_+" & "δ_-" can be calculated; according to formula (5) & (6), stability of furnace temperature "Δt_+" & "Δt_-" can be calculated as below:

$$\Delta\theta_+ = t_{pmax} - t_C \tag{1}$$

$$\Delta\theta_- = t_{pmin} - t_C \tag{2}$$

Where,

t_{pmax} refers to maximum temperature value for each measurement position (unit: °C);

t_{pmin} refers to minimum temperature value for each measurement position (unit: °C);

t_C refers to actual temperature for geometric center (unit: °C);

$$\delta_+ = t_A - t_{C'} \tag{3}$$

$$\delta_- = t_i - t_{C'} \tag{4}$$

Where,

$t_{C'}$ refers to average of indication values for geometric center (unit: °C);

t_A refers to maximum value larger than $t_{C'}$ measured at geometric center (unit: °C);

t_i refers to minimum value lower than t_i measured at geometric center (unit: °C);

$$\Delta t_+ = t_{pmax} - t_b \tag{5}$$

$$\Delta t_- = t_{pmin} - t_b \tag{6}$$

Where,

t_b refers to nominal temperature (unit: °C);

The maximum temperature difference in furnace is calculated through subtracting minimum value by maximum value measured at each temperature measurement position during each measurement cycle. The measured pressure value is indicated with average value and tightness of autoclave is determined upon selected pressure fluctuation within 5min. [12] See Table 3 for results.

Table 3. Performance results for thermal field of autoclave

Uniformity of furnace temperature /°C		Stability of furnace temperature /°C		Deviation of furnace temperature /°C		Maximum temperature difference in furnace /°C	Measured pressure value /kPa	Tightness of tank /kPa
$\Delta\theta_+$	$\Delta\theta_-$	δ_+	δ_-	Δt_+	Δt_-			
0.6	0.0	0.1	-0.1	1.0	0.2	0.8	-0.9	10.0

6. Conclusions

According to above calibration results, for distribution of thermal field of such autoclave, temperatures of top layer and bottom layer are higher than middle layer and temperature of top layer is the highest. For front, middle and rear test faces, temperature if highest for front face near bottom of autoclave and lowest for face near the door. The uniformity and deviation of furnace temperature comply with requirements; stability of furnace temperature is good; tightness and pressure values also vary in reasonable range.

References

[1] Zhang G Y Improvement of sealing structure of solidified tank *Lubrication And Sealing.* (1996)6:43

[2] Wang J Z and Feng R F Application of Furnace Tracker in Vacuum *Measurement & control technology.* (2011)30:318-320

[3] Xu H Q and Xu H T Investigation on High Stability Largescale Air Temperature Calibration Test Box *Metrology & Measurement Technology.* (2018)6:15-18

[4] Yang J J Mei X J and Yu H Y On-line measurement of vacuum furnace temperature *Metrology & Measurement Technology.* (2017)15:111-112

[5] TMI Orion French company TMI Orion has launched the world's smallest automatic radio digital recorder *Chinese instrument.* (2005)15:233-235

[6] Dong X H Xu Y M and Li Q National standard GB/T 9425—2012 "Testing method for work zone of heat treatment furnace"explanation and practice *HEAT TREATMENT OF METALS.* (2015)40:214-220

[7] Wang Y F Feng Z Q and Zhang Y Y Application of ANSYS in temperature field modeling of vacuum furnace *Metal heat treatment.* (2017)6:214-220

[8] Hua L L An L L and Kuang H H Analysis of Temperature Field Uniformity of Composite Mold in Autoclav *Journal of Nanjing University of Aeronautics & Astronautics.* (2019)51:358-364

[9] Xue W L Gao S and Wang Y Vacuum heat treatment technology for superalloys used in aero-engine *HEAT TREATMENT OF METALS.* (2019)44:12-17

[10] Song X M and Liu R X Application and Calibration of SMT Temperature Tracker in Coating and Painting *Modern Finishing.* (2015)3:49-54

[11] Wang J Han M L and Deng L W Detection and application of vacuum quenching furnace *HEAT TREATMENT OF METALS.* (2019)44:228-231

[12] He Q Yu W and Chen Y Study on temperature and pressure control of autoclavebased on FPGA *Manufacturing Technology & Machine Tool.* (2019)8:58-62

Longitudinal Protection of LCC-VSC Bipolar HVDC Transmission Line Based on Model Identification and Current Polarity

NIU Dan[1], GAO Shuping[1], SONG Guobing[2], ZHANG Chu[1] and YE Huanfei[1]

[1]Xi'an University of Science and Technology, Xi'an 710054, China
[2]Xi'an Jiaotong University, Xi'an 710049, China

Email: nd1007126311@163.com

Abstract. For hybrid bipolar DC transmission systems, combined with model identification and current polarity, a new principle of longitudinal protection of transmission lines is proposed. This paper analyzes the fault network and finds: when the front end of the inverter side fails, the fault model is equivalent to the C model, and when the back side of the inverter side fails, the fault model is equivalent to the RL model; when the front end of the rectification side fails, the rectification side fault current has a positive polarity, and when the rectification side back side fails, the rectification side fault current has a negative polarity. In conclusion, DC line area failure, namely the inverter side front and rectifier side front fault at the same time, the fault model matching C model, there was a positive polarity fault current at the same time; Failure occurs outside the DC line area, namely the inverter side front and rectifier side front no fault at the same time, inverter side failure model does not match the C model, or rectifier side fault current with a negative polarity. This new principle compensates for primary protection and can be used as an improvement to backup protection.

1. Introduction

The phase-controlled converter-type direct current (LCC-HVDC) transmission technology has many drawbacks. For example, the problem of commutation failure is inevitable, and a large amount of reactive power compensation is required, and it cannot work under the extremely weak AC system of the receiving end. The voltage source type direct current transmission (VSC-HVDC) based on the fully controlled device compensates for the defects of the LCC-HVDC, but there are also disadvantages. Combining LCC and VSC HVDC transmission technology can complement each other and maximize the advantages of both. Therefore, hybrid research on DC transmission has become a hot issue.

Literature [1-2] proves that the key component of HVDC transmission technology is the line. Therefore, studying its protection principle is of great value. It can be known from the literature [3] that the traveling wave protection method has always occupied a large proportion in the line protection, but the anti-interference and reliability are poor. The protection principle of parameter identification proposed in literature [4] shows advantages in reliability and quick movement. Literature [5] applies parameter identification in the frequency domain, while literature [6] applies parameter identification in the time domain. In general, the higher level of parameter identification problem of complex, lower

Content from this work may be used under the terms of the Creative Commons Attribution 3.0 licence. Any further distribution of this work must maintain attribution to the author(s) and the title of the work, journal citation and DOI.
Published under licence by IOP Publishing Ltd

movement speed then. To solve this problem, the protection of the principle of pattern recognition can solve [7-10], it transforms the problem of fault diagnosis into a comparison problem of model error.

In this paper, the LCC-VSC bipolar system is used as the analysis object. Combined with the model identification and current polarity, a new longitudinal protection principle of LCC-VSC bipolar line based on model identification and current polarity is proposed.

2. LCC - VSC bipolar HVDC system structure

The structure of the LCC-VSC bipolar hybrid HVDC system is shown in figure 1. The left and right ends are the rectifier station and the inverter station, which are denoted as the M side and the N side, respectively. Such wiring shown in figure 1 is applied to a high voltage, large capacity and high reliability transmission conditions [11]. The positive and negative poles of the VSC side are connected in parallel with a large capacitor.

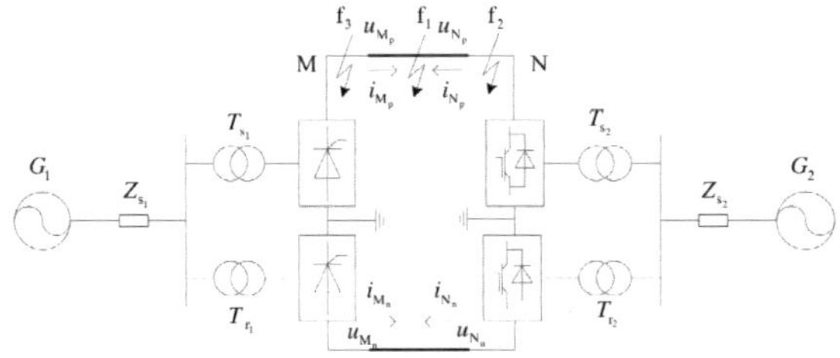

Figure 1. Bipolar hybrid DC transmission system.

In figure 1, G and Z_s represent the equivalent AC system and system impedance. Wherein, subscript 1 represents the rectification side, and subscript 2 represents the inverter side; i_M and u_M represent the current and voltage of the rectifier, i_N and u_N represent the current and voltage of the inverter side. Wherein, the subscript p represents the positive pole, and the subscript n represents the negative pole; f indicates where the fault occurred. Wherein, the subscript 1 represents the inside of the DC transmission line, the subscript 2 represents the outside of the rectifying side flat-wave reactor, and the subscript 3 represents the outside of the inverter side capacitor. The reference direction is shown in figure 1.

3. Fault Characteristics Analysis of LCC-VSC HVDC Line

3.1. External fault of the DC line

The network with a line external fault on the N side and M side of the DC line is shown in figure 2 and figure 3. u_f is the fault voltage source, i_f is the fault current, and C_p is the large capacitor; *L, R, C* are the equivalent parameters of the line, R_s and L_s are the equivalent impedance parameters of the system. Among them, the subscripts M and N represent the rectifier side and the inverter side. The line π model is adopted. The current reference direction is indicated in the figure.

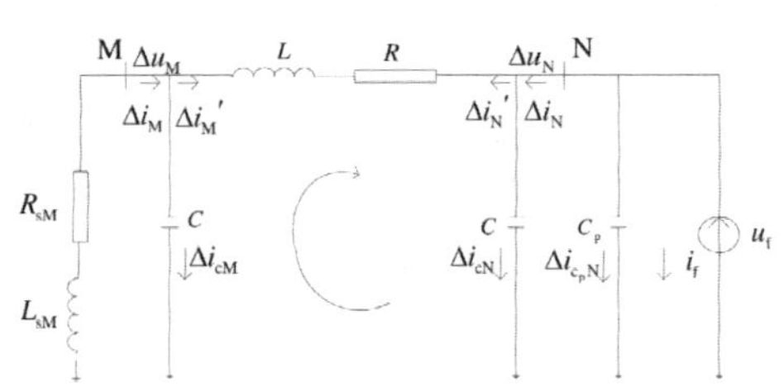

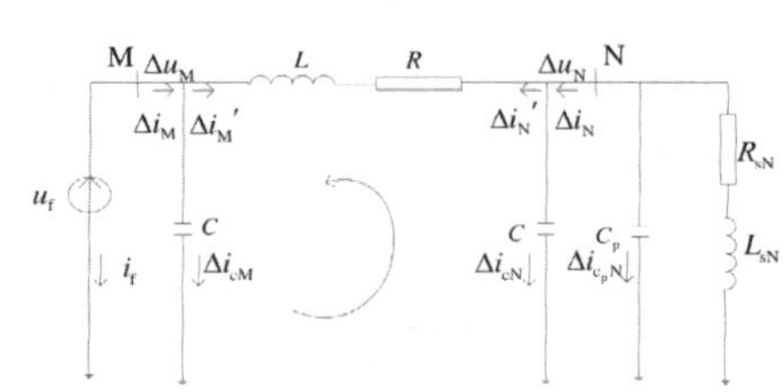

Figure 2. N-side external fault additional network. **Figure 3.** M-side external fault additional network.

3.1.1. DC line external fault on the N side. As shown in figure 2 above, when a fault occurs outside the N-side capacitor, on the M side, the rectifier side fault current has a positive polarity due to the fault voltage source [12]. Then there are:

$$\Delta i_{\mathrm{M}} > 0 \tag{1}$$

On the N side, there is:

$$\Delta u_{\mathrm{N}} = (R_{\mathrm{sM}} + R)\Delta i_{\mathrm{N}} + (L_{\mathrm{sM}} + L)\frac{\mathrm{d}\Delta i_{\mathrm{N}}}{\mathrm{d}t} \tag{2}$$

Equation (2) can be explained that by using the single-ended amount on the N side, the N-side external fault model can be equivalent to the C model.

3.1.2. DC line external fault on the M side. As shown in figure 3 above. When a fault occurs outside the flat-wave reactor on the M side, for the M side, the rectifier side fault current has a negative polarity due to the fault voltage source. Then there are:

$$\Delta i_{\mathrm{M}} < 0 \tag{3}$$

On the N side, the converter shunt Δi_{sN} and the line reactance shunt Δi_{cN} are ignored. It follows that:

$$-\Delta i_{\mathrm{N}} = C_{\mathrm{p}}\frac{\mathrm{d}\Delta u_{\mathrm{N}}}{\mathrm{d}t} \tag{4}$$

Equation (4) can be explained that by using the single-ended amount on the N side, the M-side external fault model can be equivalent to the C model.

3.2. DC line internal fault
The fault network diagram in the transmission line area in Figure 1 is shown in Figure 4.

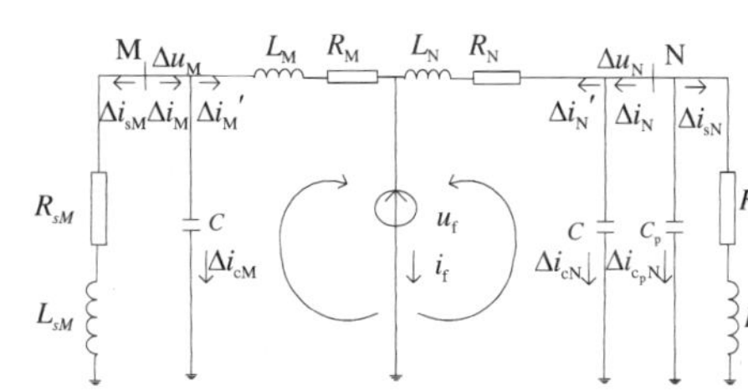

Figure 4. Additional network for internal line faults.

On the M side, the rectifier side fault current has a positive polarity due to the fault voltage source. which is:

$$\Delta i_{\mathrm{M}} > 0 \tag{5}$$

On the N side, there is:

$$-\Delta i_{\mathrm{N}} = C_{\mathrm{p}}\frac{\mathrm{d}\Delta u_{\mathrm{N}}}{\mathrm{d}t} \tag{6}$$

Equation (6) can show that the line internal fault model can be equivalent to the C model.

4. Protection principle and criteria

4.1. Model identification protection principle and criteria
It can be seen from the analysis in Section 2 that using the single-ended quantity on the N side, the fault model of the M side external and the internal line can be equivalent to the C model; the N side external fault model is equivalent to the RL model. To distinguish between the two models, define the capacitance model error function as follows:

IOP Conf. Series: Materials Science and Engineering **752** (2020) 012021 doi:10.1088/1757-899X/752/1/012021

$$E_{\mathrm{C}}=\frac{\int_0^T\left|\Delta i_{\mathrm{N}}-C_{\mathrm{p}}\frac{\mathrm{d}\Delta u_{\mathrm{N}}}{\mathrm{d}t}\right|\mathrm{d}t}{\int_0^T\left(\left|\Delta i_{\mathrm{N}}\right|+\left|C_{\mathrm{p}}\frac{\mathrm{d}\Delta u_{N}}{\mathrm{d}t}\right|\right)\mathrm{d}t} \tag{7}$$

In equation (7): T=10ms.

The construction protection criteria are as follows:

$$E_{\mathrm{C}}<\mu_{\mathrm{set}} \tag{8}$$

In equation (8): μ_{set} is the capacitance model error setting value.

4.2. Protection principle and criterion using current polarity

It can be seen from the analysis in Section 2 that using the fault current symbol on the M side, it can be known that $\Delta i_M<0$ on the M side external fault; $\Delta i_M>0$ in the line internal fault and the N side external fault. Thus, a fault current average function is constructed to reflect the polarity of the fault current.

The function of defining the fault current average is as follows:

$$i=\frac{1}{N}\int_0^T\Delta i\mathrm{d}t \tag{9}$$

In equation (9): T = 10ms, where N is the corresponding number of sampling points.

Construction protection criterion is as follows:

$$\frac{1}{N}\int_0^T\Delta i\mathrm{d}t>k*I_{\mathrm{set}} \tag{10}$$

In equation (10) : T=10ms, N is the corresponding number of sampling points.k is the reliability factor, 1.2~1.5 is acceptable; $I_{\mathrm{set}}=0.1*I_n$, I_n is the rated current value of the DC system.

4.3. Protection criteria implementation method

According to the analysis in sections 4.1 and 4.2, the capacitance model starts when the error is less than the threshold value, and starts when the current mutation component is greater than the threshold value. The constructor is as follows:

$$F(A,B)=A\mathrm{and}B \tag{11}$$

$$A=sign(\mu_{\mathrm{set}}-E_{\mathrm{C}}) \tag{12}$$

$$B=sign(\frac{1}{N}\int_0^T\Delta i\mathrm{d}t-k*I_{\mathrm{set}}) \tag{13}$$

In equation (12) and equation (13): $sign(\bullet)$ is a sign function, when $(\bullet)>0$, $sign(\bullet)=1$; when $(\bullet)<0$, $sign(\bullet)=0$. T=10ms, where N is the corresponding number of sampling points.

It can be seen that when $F(A,B)=1$, the capacitance model error is small at this time, the polarity of the rectifier side fault current is positive, and it is judged as the internal fault of the line; when $F(A,B)=0$, it is judged as the external fault of the line.

5. Current polarity and model identification longitudinal protection strategy

When the capacitance model error is large, it is determined as the back-side fault on the N side, and the fault occurs outside the HVDC transmission line. At this point, the N side does not need to send fault signals to the M side. Only when the fault occurs in front of the N side, i.e. the capacitance model error is small, will the fault signal be sent to the M end, and the capacitor model error discriminating element on the inverter side is started.

When the fault occurs behind side M, the fault current at side M has negative polarity, i.e. $\Delta i_M < 0$. At this time, it is not necessary to start the discriminating element of current polarity. The discriminating element is activated only when the fault occurs in front of the M side, that is, $\Delta i_M > 0$. In short, only when the current polarity discriminating element on the M side is activated and the feedback of the N side capacitive model error discriminating element action information is received, it can be known that the fault occurs inside the DC line, the protection device operates, and the reliability is relatively high. The logic diagram of the longitudinal protection action is shown in figure 5.

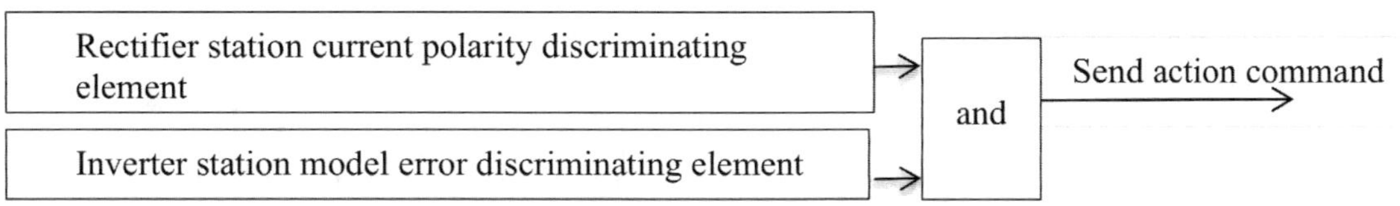

Figure 5. Principle of longitudinal protection.

As can be seen from figure 5, the protection device transmits an action command only when the two-stage converter station discriminating elements operate.

6. Simulation

The simulation model of the 500 kV LCC-VSC hybrid bipolar HVDC system is shown in figure 1. The rated current is 2 kA and the total length of the DC line is 1000 km. The capacitance of the positive and negative poles in parallel on the VSC side is 1000 μF, and the data acquisition frequency is 10kHz. The fault occurred at 1.85 s and lasted for 0.1 s. The data window takes 10ms and the sliding window takes 20ms. To ensure reliability, μ_{set} is taken as 0.5.

6.1. DC line external fault simulation results

6.1.1. DC line external fault simulation results on the N side. When a metallic grounding occurs outside the capacitor on the N side, the fault simulation diagram is shown in figure 6. (corresponding to f_2 in figure 1)

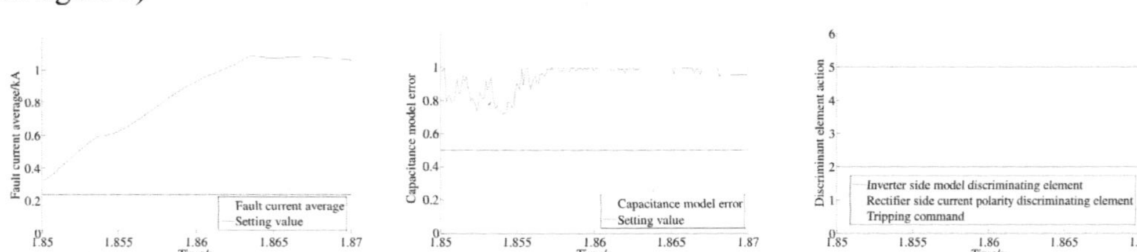

(a) (b) (c)

Figure 6. Metal ground fault occurs outside the DC line on the N side.

As can be seen from figure 6, the outside of the DC line on the N side is faulty. Within 5 ms~15 ms after the fault, the calculated average value of the rectifier side fault current is greater than the set value $k * I_{set}$, but the error value of the inverter side capacitor model is greater than the set value μ_{set}. At this time, the current polarity discriminating element is activated, but the capacitance model discriminating element is not activated. It is judged to be an external fault of the DC line and the protection device is not activated.

IOP Conf. Series: Materials Science and Engineering **752** (2020) 012021 doi:10.1088/1757-899X/752/1/012021

6.1.2. DC line external fault simulation results on the M side. When metal grounding occurs outside the flat-wave reactor on the M side, the fault simulation diagram is shown in figure 7. (corresponding to f_3 in figure 1)

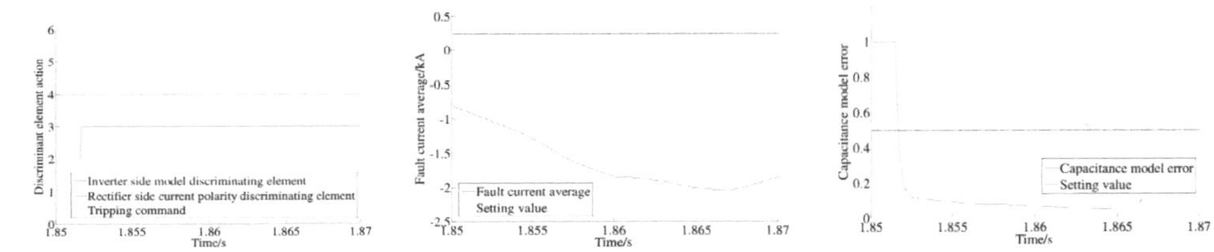

(a) (b) (c)

Figure 7. Metal ground fault occurs outside the M side.

As can be seen from figure 7, the outside of the DC line on the M side is faulty. Within 5 ms~15 ms after the fault, the calculated error value of the inverter side capacitor model is less than the set value μ_{set}, but the average value of the rectifier side fault current is less than the set value $k * I_{set}$. At this time, the capacitance model discriminating element is activated, but the current polarity discriminating element is not activated, so it is determined that the external fault of the DC line does not activate the protection device.

6.2. DC line internal fault simulation results

When a metal ground is generated at a position closer to the N side inside the DC line, the fault simulation diagram is shown in figure 8; When the high-resistance grounding occurs at a position closer to the M side, the fault simulation diagram is shown in figure 9. (corresponding to f_1 in figure 1)

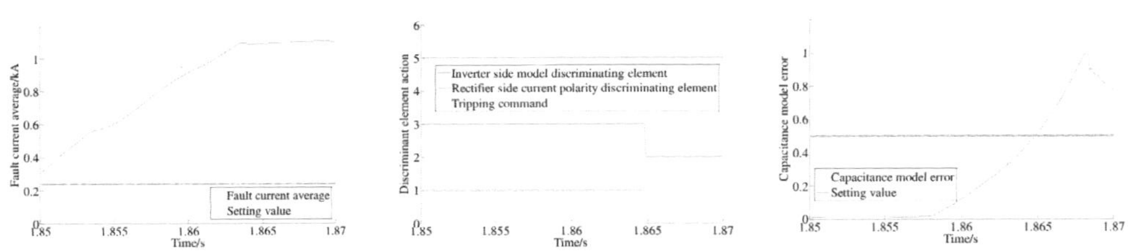

(a) (b) (c)

Figure 8. Metal grounding 50 km from the N end.

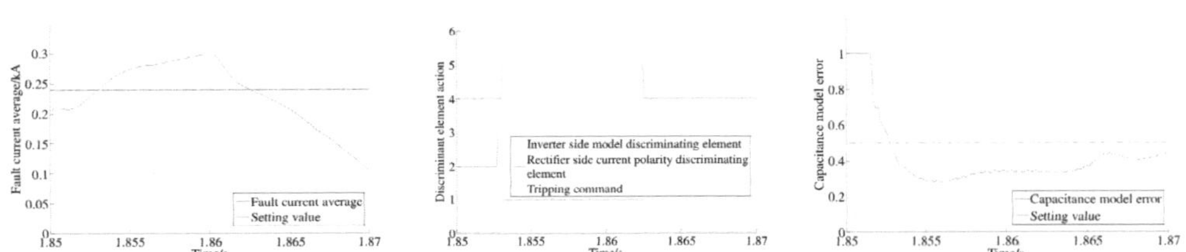

(a) (b) (c)

Figure 9. Grounded through a 500 Ω transition resistor at 50km from the M end.

It can be seen from figure 8 to figure 9 that different ground faults occur at different positions inside the DC line. Within 5ms to 15ms after the fault occurs, the calculated average value of the rectifier side fault current is greater than the set value $k*I_{set}$, and the capacitance model error value on the inverter side is less than the set value μ_{set}. At this time, the current polarity discriminating element is activated, and the capacitance model discriminating element is also activated. It is determined that the internal fault of the line and the protection device is activated.

It can be seen from the above simulation that the simulation results of the capacitance model discriminating element on the inverter side and the current polarity discriminating element on the rectifying side are consistent with the theoretical analysis.

7. Conclusion

Aiming at the LCC-VSC hybrid bipolar DC transmission system, a new principle of longitudinal protection for DC transmission lines based on model identification and current polarity is proposed. Analysis and simulation show that this principle can respond to faults 5 to 15ms after a fault, without filtering, resistance to transition resistance, simple and easy adjustment. It can compensate for the primary protection or it can be used as an improvement to the backup protection.

Acknowledgments

This work is supported by National Natural Science Foundation Joint Fund Key Support Project (No. U1766209) and National Natural Science Foundation of China (No. 51777166).

References

[1] Zhao Yuejiao. Fault detection of HVDC transmission line based on improved Blackman window [D]. *Northeast Petroleum University*, 2018.
[2] Zhai Xingfu, Song Guobing, Xu Haiyang, et al. New Principle of Pilot Protection for VSC-HVDC Transmission Lines Using Modulus Model Identification[J]. *Automation of Electric Power Systems*, 2014, (10): 100-106.
[3] Qi Guoqiang, Wang Zengping. Hilbert-Huang Transform Based HVDC Mutation Direction Pilot Protection Method[J]. *Power System Protection and Control*, 2017, 45(20): 92-99.
[4] Suo Nanjiale,Kang Xiaoning,Song Guobing, et al.Preliminary study on the principle of relay protection based on parameter identification[J].*Journal of Electric Power System and Automation*,2007,19(1):14-20,27.
[5] Yang Lijun, Peng Pan, Gao Wei, et al. Interval identification of moisture and aging of oil-paper insulation based on frequency domain dielectric response signature fingerprints[J]. *Transactions of China Electrotechnical Society*, 2018,33(9):2105-2114.
[6] Hou Junjie,Fan Yanfang,Wang Yibo.Research on Time Doman iDistance Protection of Parameter Identification for Cluster Wind Power Transmission Line[J].*Power System Protection and Control*,2018,46(5):46-53.
[7] SONG Guobing,YAN Xingfu,LI Dekun, et al.Vertical protection of voltage source converter type DC transmission line for identifying model parameters[J].*Journal of Xi\'an Jiaotong University*,2013,47(8):110-114.
[8] Zhai Xingfu, Wu Xiaoming, Zheng Jiafa, Zhang Jinfeng, Zhu Can. Longitudinal Protection of HVDC Transmission Lines Based on Model Identification[J]. *Electric Power Survey and Design*, 2017(S2): 202-206+335.
[9] SONG Guobing,YAN Xingfu,YAN Mengbing, et al. Pilot protection of VSC-HVDC transmission line based on parallel capacitance parameter identification[J].*Automation of Electric Power Systems*,2013,37(15):76-82,102.
[10] Yuan Hailong, Yuan Yuan, Song Hailian. A new method for longitudinal protection of HVDC transmission lines [C]. *//2013 Proceedings of the Annual Meeting of China Electrical Engineering Society*. 2013:1-6.
[11] Zhao Chengyong, Guo Chunyi, Liu Wenjing. *Mixed DC Transmission* [M].2014
[12] Gao Shuping, Sonan Jiale, Song Guobing, et al. New Principle of Longitudinal Protection of HVDC Transmission Lines Using Current Mutation Characteristics[J]. *Automation of Electric Power Systems*, 2011, 35(5): 52-56, 86.

An Alternative Method for GIS Insulation Performance Evaluation

Song Lin Wu[1], Bo Xiao[1], Min Hu[2,3], Yong Yao[1], Wei Fu[1] and Haiyan Zhang[2]

[1]State Grid Chongqing Electric Power Company Urban Power Supply Branch,Chongqing, China
[2]Chongqing University of Science & Technology, China

174661848@qq.com

Abstract. The continuous improvement of system capacity and voltage level is somewhat accompanied by insulation fault as one of the main faults of GIS device. In this paper, an alternative method for identifying the GIS insulation performance based on fuzzy evidence theory is proposed by way of judging the four main features of GIS insulation level. Firstly, the fuzzy statistic test was used to obtain the normalized values of the four feature quantities of GIS insulation level. Four methods were used to calculate their respective expectations and deviations. The membership degree value was determined by the membership function, and the difference between the membership degrees was also considered. The degree of fusion was used to determine the weight of each feature quantity, so that the membership degree of the same kind of feature information fusion was obtained. Finally, the merged membership degree was transformed into the basic trust distribution function. The evidence theory then took the job to realize the pairwise fusion of the trust distribution functions between different feature quantities. The experimental results showed that this alternative way was proven to make effective judgment on the GIS insulation performance.

1. Introduction

The development of computer technology is accompanied by the improvement of artificial intelligence and testing technology. In GIS device, the condition monitoring circuit breaker is oriented by the large-scale expert system based on intelligent theory and neural network. At present, there are some results on the research of equipment state assessment at home and abroad, but it remains difficult to perform the study on the state assessment of GIS. As we all know, the research work on GIS device status assessment is not only of important academic significance and social benefits, but also has shown impressive values in economic and engineering applications. More investments in GIS state assessment is therefore encouraged so as to continuously improve the related technologies, which would then be beneficially promoted in the power system.

GIS device failures can be roughly divided into two categories: discharge faults and mechanical faults. Specifically, major faults include insulation faults, mechanical faults, casing leaks, and secondary loop faults. Looking to the previous investigation and statistics, there are five main reasons that are responsible for the occurrence of GIS insulation faults (as shown in figure 1). Among them, SF6, which is an indispensable insulating gas for GIS device, will decompose under the action of arc, electric spark and corona discharge, resulting in insulation degradation. This situation may be complicated by SF6 gas leakage, increase of water content, decomposition of the gas composition, etc. As a result of the impact

Content from this work may be used under the terms of the Creative Commons Attribution 3.0 licence. Any further distribution of this work must maintain attribution to the author(s) and the title of the work, journal citation and DOI.
Published under licence by IOP Publishing Ltd

on the insulation strength of GIS device, the probability of insulation accidents accounted for 38.1% of GIS accidents. Obviously, it does make sense to evaluate the insulation state of GIS.

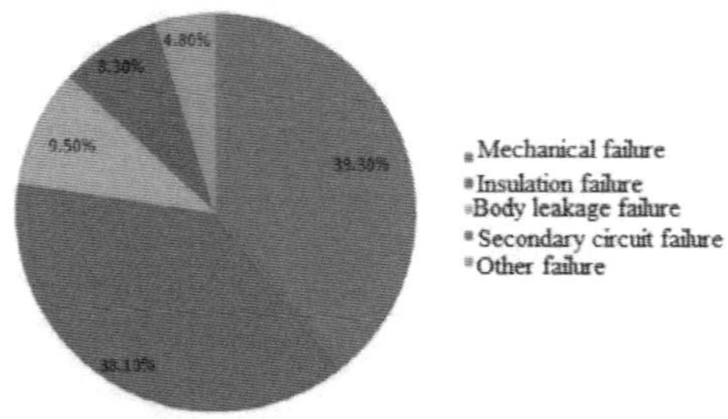

Figure 1. Failure Rate on Common Fault Types of GIS Device.

Huang Yucheng et al. in Jiangsu Tianfang Power Technology Co., Ltd. introdued DR detection technology to achieve visual diagnosis of the internal structure of GIS device, being able to effectively diagnosing the internal defects of GIS [1]. Lu Fangcheng et al., who worked in the key laboratory for safety and defense of power transmission and transformation equipment in North China Electric Power University, used ultrasonic sensors to collect waveforms under three discharge models. The collected data were identified by way of BP neural network. The recognition rate was app. 80% [2]. Wang Taoyun et al., from the School of Energy and Electric Engineering, Hohai University, combined the Petri net theory with fuzzy inference rules. With reference to a large number of statistical fault cases, the correctness of the methodology was verified by establishing a diagnostic model to find out the cause of the fault in the process of reverse fuzzy reasoning. Liu Yongye et al. took the feature gas as the input library of the fuzzy Petri net, and established a diagnosis model for GIS insulation fault. This made it possible for the Petri net to analyze the credibility of the GIS defect type [3-4]; Yan Xianglian from China National Electric Power Research Institute tested the decomposition products of SF6 and obtained the decomposition product features of equipment, thereby judging the operating conditions of SF6 equipment in different types[5] .Wang Taoyun, who worked with State Grid Shanghai Electric Power Company Jinshan Power Supply Company, introduced the extension analysis theory and solved the weight problem of each index in GIS state assessment with the aid of entropy method theory, which accurately evaluated the gas chamber status of GIS[6].

2. Membership

Monitoring the GIS evaluation parameter is mainly to check the state of SF6 gas, including (1) SF6 gas detection, (2) resistance parameter, (3) accessory parameter analysis, (4) working environment and other factors parameter analysis, wherein the SF6 gas detection is directed to the trace water content of the SF6 gas; the pressure of the SF6 gas; the density of the SF6 gas; the composition of the SF6 gas. The GIS insulation state is mainly related to the SF6 gas state, so the gas trace water content, the arc extinguishing chamber including other gas chambers, the gas leakage rate and the partial discharge product were used in this work as the four feature quantities for evaluating the GIS insulation performance.

The four feature quantities of GIS insulation performance were determined by means of fuzzy statistical test.

As for these four feature quantities of GIS, the smaller the better, so choose formula (1) as the normalization function.

$$f(x) = exp\,((x - b)/b) \tag{1}$$

Where: f(x) is the membership function argument, x is the parameter raw data, and b is the threshold specified in the procedure.

According to the State Grid Corporation's substation testing and management regulations, in the State Grid (Yunjian/3) 829-2017, the standard for SF6 gas testing items stipulates that trace water content in the operation of the compartment with arc decomposition is required not exceed 300μL/L, the

IOP Conf. Series: Materials Science and Engineering **752** (2020) 012022 doi:10.1088/1757-899X/752/1/012022

compartment free of arc decomposition shall not exceed 500 μL/L; the gas leakage rate shall not exceed 1%; the partial discharge (PD) product shall not exceed 10 μL/L; thus, the threshold values of all feature quantities are known. A total of 28 experts were invited to fill out the questionnaire to determine the degree of trace water membership in the arc chamber, as shown in table 1.

Table 1. State Evaluated by Experts Corresponding to the Trace Water Content of the GIS Arc Extinguishing Chamber.

Insulation state	Good	Average	Note	Danger
1	70	150	230	265
2	120	200	220	240
⋮	⋮	⋮	⋮	⋮
28	100	120	160	245
Range	0-170	120-224	155-268	192-300

The corresponding normalized values of Table 1 were obtained from Equation 1, and the other gas chamber had the similar trace water content, gas leakage rate and PD products.

The GIS insulation state boundary is not particularly obvious, so the different insulation states of GIS can be described by information entropy. Information entropy describes the degree of chaos in the system. The more chaotic the system is, the higher the entropy is. Different entropies can be described in terms of the order of different aspects. The system information entropy is defined as: Let W be an algebra generated by the measurable set class v and a Lebes space with p measure, p(W) =1, and W can be expressed in its finite partition C = {Ci} as the form of mutually incompatible collections, i.e.:

$$W = \cup_{i=1}^{n} C_i 且 C_i \cap C_j = \emptyset, \forall i \neq j \tag{2}$$

Then the information entropy for this partition C:

$$H(C) = -\sum_{i=1}^{n} p(C_i) log(C_i) \tag{3}$$

In this work, four entropy features including wavelet energy spectrum entropy, power spectral entropy, singular spectral entropy and wavelet spatial state feature entropy were used to describe the GIS insulation state. The same entropy feature of GIS various insulation cases also obeys Gaussian distribution. Hence, the membership function was chosen as Gaussian type, i.e.

$$\mu(x) = exp(-\frac{(x-a)^2}{b^2}) \tag{4}$$

Where, a, b are the mathematical expectation and the mean square error, so that the membership function matrix is a goal within reach. The following table 2-5 present the corresponding a and b values among various entropy values of the GIS feature quantity.

Table 2. Mathematical Expectation and Mean Square Error of Wavelet Energy Spectral Entropy in Different Cases.

Feature quantity		Good	Average	Note	Danger
Trace water content in arc chamber	a	0.8447	0.8453	0.8447	0.8450
	b	0.0009	0.0004	0.0006	0.0002
Trace water content in other air chambers	a	0.8446	0.8449	0.8452	0.8449
	b	0.0005	0.0004	0.0002	0.0003
Gas leakage rate	a	0.8315	0.8446	0.8437	0.8446
	b	0.0120	0.0039	0.0029	0.0004
PD product	a	0.8451	0.8452	0.8450	0.8451
	b	0.0001	0.0002	0.0003	0.0001

IOP Conf. Series: Materials Science and Engineering **752** (2020) 012022 doi:10.1088/1757-899X/752/1/012022

Table 3. Mathematical Expectation and Mean Square of Power Spectrum Entropy in Different Cases.

Feature quantity		Good	Average	Note	Danger
Trace water content in arc chamber	a	0.4247	0.4213	0.4332	0.4217
	b	0.0572	0.0520	0.0408	0.0024
Trace water content in other air chamber	a	0.3823	0.4194	0.4235	0.4395
	b	0.0296	0.0059	0.0318	0.0149
Gas leakage rate	a	0.3696	0.3894	0.3958	0.4139
	b	0.0358	0.0221	0.0830	0.0079
PD product	a	0.4272	0.4167	0.4373	0.4289
	b	0.0100	0.0281	0.0096	0.0092

Table 4. Mathematical Expectation and Mean Square of Singular Spectrum Entropy in Different Cases.

Feature quantity		Good	Average	Note	Danger
Trace water content in arc chamber	a	0.1140	0.1038	0.0991	0.0869
	b	0.0022	0.0001	0.0205	0.0311
Trace water content in other air chamber	a	0.1089	0.0755	0.0705	0.0621
	b	0.0152	0.0131	0.0352	0.0052
Gas leakage rate	a	0.2443	0.2092	0.1690	0.1204
	b	0.0433	0.0020	0.0085	0.0336
PD product	a	0.0519	0.1136	0.0507	0.0555
	b	0.0163	0.0060	0.0054	0.0085

Table 5. Mathematical Expectation and Mean Square Error of Wavelet Spatial State Feature Entropy in Different Cases.

Feature quantity		Good	Average	Note	Danger
Trace water content in arc chamber	a	0.0939	0.0711	0.0578	0.1033
	b	0.0743	0.0536	0.0177	0.0415
Trace water content in other air chamber	a	0.0661	0.0504	0.0503	0.0519
	b	0.0334	0.0157	0.0240	0.0064
Gas leakage rate	a	0.2242	0.1964	0.0999	0.0718
	b	0.0472	0.0573	0.0543	0.0793
PD product	a	0.0559	0.1264	0.0600	0.0430
	b	0.0494	0.0216	0.0181	0.0267

3. Discrimination of insulation state

After extracting four kinds of entropy features from the sample data of four feature quantities, the parameters of the membership function were determined to perform data fusion of different feature quantities. For example, extract the sample value of the trace water content of the arc extinguishing chamber for a certain period of time, and obtain the wavelet energy spectrum values corresponding to the sample, power spectrum, singular spectrum, and wavelet spatial spectrum. They were substituted into the formula (4) to get the values of insulation state memberships:

$$\mu = \begin{bmatrix} 0.9965 & 0.5283 & 0.9863 & 0.7108 \\ 0.7175 & 0.5744 & 0.8122 & 0.3950 \\ 0 & 0.2522 & 0.6987 & 1 \\ 0.1624 & 0.3109 & 0.0868 & 0.5650 \end{bmatrix}$$

The degree of mutual support between the degrees of entropy was used to measure the degree of mutual support between the entropy values. The definition:

$$d_{ij} = \sqrt{\sum_{k=1}^{K}(\mu_{ik}(x_i) - \mu_{jk}(x_j))^2\,)} \quad (5)$$

the greater d_{ij}, the lower the support between the two different entropy values, and the smaller the value, the higher the support. Therefore, the fusion function of different algorithms with the same feature quantity is defined:

$$r_{ij} = 1 - \sqrt{\sum_{k=1}^{k}(\mu_{ik}(x_i) - \mu_{jk}(x_j))^2\,)} \quad (6)$$

Get the maturity matrix:

$$r = \begin{bmatrix} 1.0000 & 0.7900 & 0.2354 & 0.5732 \\ 0.7900 & 1.0000 & 0.0026 & 0.0674 \\ 0.2354 & 0.0026 & 1.0000 & 0.4066 \\ 0.5732 & 0.0674 & 0.4066 & 1.0000 \end{bmatrix}$$

To ensure the maximum credibility, when determining the fusion degree of the i-th entropy value with other entropy values, the smallest of the other entropy values and the i-th entropy value may be obtained, that is,

$$s = [0.2354 \quad 0.0026 \quad 0.0026 \quad 0.0674]^T$$

Let the weight coefficient of the i-th feature quantity be determined by s_i in the proportion of the total phase fusion, i.e.:

$$q_i = \frac{s_i}{\sum_{i=1}^{N} s_i} \quad (7)$$

The resulting weight coefficient vector:

$$q = [0.7641 \quad 0.0085 \quad 0.0085 \quad 0.2188]^T$$

After fusion, the degree of membership of the arc extinguishing chamber trace water content corresponding to each insulation state:

$$\beta = [0.7615 \quad 0.0045 \quad 0.0084 \quad 0.1555]$$

From the above, the fusion of the insulation performance evaluation of a certain feature quantity is completed, and the result of discriminating the insulation state of the other three feature quantities can be obtained by analogy. Then it's the turn of the evidence theory to do its part on insulation discrimination. The evidence theory works well with uncertain information, and it has been widely used in fault diagnosis detection, multi-objective recognition, information fusion, and uncertainty multi-attribute decision-making. Specifically, the acquisition of the mass function is the key to the application of evidence theory [7].Han Feng [8] et al. introduced a method of credibility distance; Xu Yanke et al. [12] explored the methods for fuzzy sets; Xu Jiali [9] counted on the information amount of evidence to determine the credibility, thus obtaining the mass function. All have achieved certain results that matter.

Since the GIS insulation performance extracted the M-class features, the corresponding basic trust distribution functions were m_1, m_2, ... m_M, and each type of feature corresponded to the same K-class insulation state. In this way, the focal point of the basic trust distribution function was the same.

The value of the basic trust assignment function was transformed from the merged membership value. To satisfy $\sum m(A_i) = 1$, it was determined by the following formula:

$$\mathrm{m_r(A_i)} = \beta_i^{(r)} / \sum_{i=1}^{K} \beta_i^{(r)} \quad (8)$$

The converted basic trust assignment function value from above:

$$m1 = [\,0.8189 \quad 0.0049 \quad 0.0091 \quad 0.1672\,]$$

Similarly, the basic trust assignment function values of other features were obtained:

$$m2 = [\,0.2412 \quad 0.3015 \quad 0.1573 \quad 0.3000]$$
$$m3 = [0.2691 \quad 0.0629 \quad 0.6426 \quad 0.0254]$$
$$m4 = [0.2463 \quad 0.2275 \quad 0.3808 \quad 0.1454\,]$$

In the presence of the many types of features, the evidence theory worked to combine the pairwise features, that is, the pairwise fusion of m1, m2, m3, and m4:

$$m = [\,0.9615 \quad 0.0016 \quad 0.0257 \quad 0.0136]$$

Judging from this, the GIS insulation type was good, which was consistent with the actual situation. The same method was applicable for identification of other sample data. They are not listed here in details.

4. Conclusion

Four feature quantities of continuous sampling GIS were analyzed for judgment with the corresponding evaluation models obtained with the aid of a large number of test data. The GIS parameters to be tested were targeted to determine the insulation capability in reference to the evaluation model. In this paper, the proposed recognition method based on fuzzy evidence theory performed well on the relationship among various features, aiming to avoid the subjectivity in the fusion process. Experiments showed the effectiveness of the proposed method in merging the GIS features in various situations.

Acknowledgments
This work was supported by State Grid Chongqing Electric Power Company Urban Power Supply (Project no. SGTYHT/18-JS-206, Contract no. SGCQSQ00BDJS1900987). The project name is State Grid Chongqing Electric Power Company Urban Power Supply 2019 Research Service of Health Assessment Model for GIS.

References

[1] Xuancheng H, Tao H and Quanyi H,2018 *J Detection of internal insulation defects of GIS devices by DR technology. Thermal Power Generation* .**47** 140-146.
[2] Fangcheng L and Bo Z. 2014 *J Identification of GIS Insulation Defect Type Based on Ultrasonic Method. Electrical Measurement and Instrumentation*. **51** 22-26.
[3] Taoyun W, Hongzhong M, Yangliu C, Ning J, Kai L and Honghua X. 2016 *J Fault Diagnosis and Reliability Analysis of GIS Based on Fuzzy Petri Nets. New Technology of Electrical Engineering and Energy*. **35(5)** 67-73.
[4] Yongye L, Hongzhong M, Yong L and Honghua X. 2018 *J GIS Fault Diagnosis Method Based on Gas Detection and Fuzzy Petri Net. Journal of Electric Power Science and Technology*. **33** 141-146.
[5] Xianglian Y, Gao S and Chengyu W. 2014 *J Gas Insulated Switchgear State Monitoring Based on SF6 Decomposition Products Detection. Electric Power Automation Equipment*.**34** 83-88, 95.
[6] Taoyun W, Hongzhong M and Yangliu C. 2016 *J Condition Evaluation of Gas Insulated Switchgear Based on Extension Analysis and Entropy Method. Power System Protection and Control*. **44** 115-120.
[7] Jousselme A L, Chunsheng L and Grenier D. 2006 *J Measuring Ambiguity in the Evidence Theory[J]. IEEE Transactions on Systems, Man and Cybernetics, Part A*.**36** 890-903 .
[8] Feng H, Wanhai Y and Xiaoguang Y. 2010 *J Evidence Theory Information Fusion Method Based on Fuzzy Set. Control and Decision*. **25** 449-452.
[9] Yanke X, Xiaogeng L and Xiaohong J.2012 *J Information Fusion Based on Fuzzy Evidence Theory and its Application in Target Recognition. JOURNAL OF HARBIN INSTITUTE OF TECHNOLOGY*. **44** 107-111.

Dynamic Electricity Price Guidance Analysis of Urban Electric Vehicle Based on Actual Data Drive

***Li Duanchao, Su Han and Wang Haiwei**

Hefei Power Supply Company, State Grid Anhui Electric Power Co., Ltd. Hefei 230088, China

*13865601314@ 163.com

Abstract. Establishing a dynamic electricity price to guide the charging behavior of car owners can better help electric vehicles (EV) to participate in peak shaving and valley filling of the grid. On the basis of this, a dynamic electricity price mechanism for electric vehicle charging stations in city was proposed in this paper.It was aimed at the distribution network where residential areas are located and mean to improve the ability of EV to cut peaks and fill valleys. Firstly, the load characteristics of residential areas were analyzed, in addition, the load status of the distribution network in the area before and after the EV charging station connected in were compared. Secondly, according to the proposed dynamic electricity price calculation process and related function based on the TOU (time-of-use) electricity price for electric vehicle charging station, the dynamic price of the day was established, then in order to predict the power load after the dynamic electricity price is implemented, the corresponding model based on the residential load elastic demand matrix was established. Finally, taking the actual data of a certain district near residential area in Anhui province and an electric vehicle charging station as an example, formulated the targeted dynamic electricity price and further verified the feasibility and effectiveness of the study.

1. Introduction

With the worldwide attention to environmental issues and the increasingly depleted energy crisis of fossil fuels, the development of clean energy is an irresistible trend. Electric vehicles use electricity instead of fuel as the power energy, which not only reduces the dependence on fossil energy, but also reduces the adverse impact on the environment. On the other hand, a large number of electric vehicle cluster loads have greater randomness in time and space compared with traditional loads, and thus have good scheduling performance. However, if a large number of electric vehicle loads are directly connected to the grid, it will bring a certain load impact to the grid. Therefore, electric vehicle clusters must be planned into the grid scheduling strategy, and use their energy storage characteristics to cut peak and fill valley.

In [1], the charging and discharging strategy of electric vehicles in micro grid was optimized for scheduling based on day-ahead load prediction combined with renewable energy generation. However, the influence of the actual electricity price on the charging and discharging behavior of electric vehicles was not considered. Study [2] considered the influence of electricity price, the charging and discharging optimization scheme for electric vehicles with the purpose of stabilizing the load curve was studied. Besides, according to the investigation, the certain realistic constraints on the interactive development of electric vehicles based on the orderly charging of electric vehicles include the lack of

Content from this work may be used under the terms of the Creative Commons Attribution 3.0 licence. Any further distribution of this work must maintain attribution to the author(s) and the title of the work, journal citation and DOI.
Published under licence by IOP Publishing Ltd

enthusiasm of electric vehicle owners for network interaction. In the traditional electric vehicle charging mode, the charging piles were occupied. The network resources were opaque, and the charging price was relatively simple. There was a lack of a business model that mobilizes the owners of electric vehicles to participate in the interaction between the network and the Netherlands. TOU is now the most widely used power business model, which can effectively motivate power users to adjust their electricity consumption behavior, especially electric vehicles which is more schedulable than ordinary users. [3] studied the charging strategy optimization of electric vehicles under TOU price. While [4] proposed an electricity pricing model that took into account the satisfaction of car owners, and optimized the solution by taking the satisfaction of users, peak and valley load difference and the lowest electricity cost as multi-objective function. However, the proposal of dynamic electricity price can further close to the characteristics of dynamic load change and more accurately affect the user's choice of electricity consumption behavior. Thus [5] proposed a dynamic pricing strategy for new energy electricity price, which was divided into two parts: fixed price and floating price. The simulation results verified the correctness of electricity price formulation.

Based on the current situation of electricity consumption in urban residential areas, this paper analyzes the residential electricity habits with actual data, the impact of electric vehicle charging on the load curve of the area according to the charging station order data is analyzed at the same time. In order to reduce grid load fluctuations, the dynamic dynamic transaction price is established based on the original TOU price to guide the orderly charging of EV and participate in grid dispatching. Then forecast the station load condition under the new electricity price mechanism with the load demand elasticity matrix. Furthermore, the status quo before and after the implementation of the new price mechanism was compared and analysed. At last prove with actual data that it is fieasible to realize the effective dispatch of electric vehicles by setting electricity prices, reduce the negative impact of electric vehicles on the power grid, and support the safe and effective operation of large power grids.

2. Impact of EV charging on residential distribution network

2.1. Analysis of residential load electricity characteristics

The general community can be divided into ordinary, mid-range and high-end residential areas according to the single-family area and household income level [6], but no matter which type of residential area, the main load of residential electrical equipment is mainly refrigerators, air conditioners, television sets, Washing machines, electric heating, water heaters, etc., which have strong electrical characteristics, are mostly seasonal loads, and because of the obvious peak-to-valley characteristics during residential electricity consumption in one day, the characteristics of electricity-polymerization response in residential areas are easy to generalize and analyze.

Taking a certain district of a certain city in Anhui Province as an example, draw the daily load curve as shown in figure 1, the typical daily load data plot is selected to find that the peak load of residential areas in one day is mainly concentrated from 19:00 to 23:00 in the evening, and two small peak period in the morning at 7:00 to 8:00 and 11:00 to 13:00, which is mainly related to the user's behavior habits.

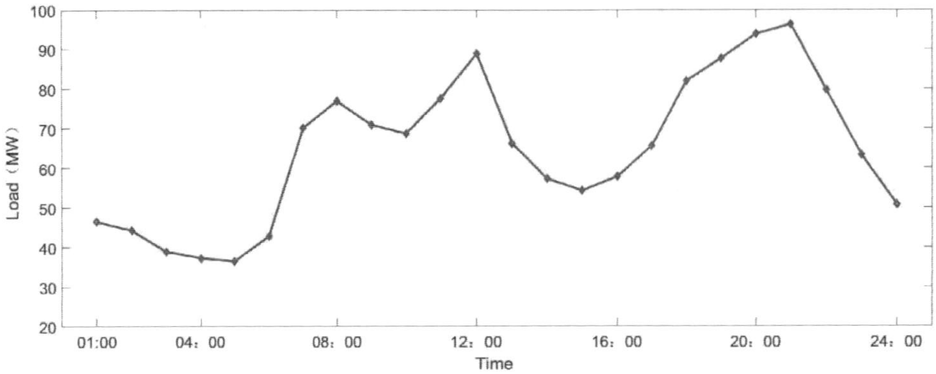

Figure 1. The daily load curve of a residential area

2.2. Influence of electric vehicle access on load characteristics

2.2.1. Electrical vehicle charging mode. The charging modes of electric vehicles on the market mainly include three modes: fast charging, slow charging and battery switch. The battery switching method has more requirements for battery parameters and interface standards, and is suitable for public transportation vehicles of a unified model. For general electric vehicle loads, there are mainly two methods: fast charging and slow charging. In the normal charging (slow charging) mode, the charging power is low while the charging time is long, thus a long parking time is required, while the fast charging can save time, but the loss of the battery is relatively large. In general large public/commercial parking lots or residential areas, users who are not sensitive to charging time are more likely to choose slow charging mode, while the driver who travels long distances or mainly based on transportation traffic prefers the choice of fast charging mode. [6]

2.2.2. Data processing of charging station. Electric vehicle charging stations are divided into two major categories, urban charging stations in cities, and electric vehicle charging stations set up on expressways. This paper mainly studies the city's conventional charging station in residential areas. Using the actual power trading order of the charging station, the charging power, charging time, and electricity price information are extracted, thus the charging characteristics of the electric vehicle under the current electricity price mechanism are analyzed.

2.2.3. Analysis of the impact of electric vehicle access on power grid. In the context of demand-side management development, the charging of electricity charges has gradually evolved from the traditional unified electricity price to the TOU price mechanism that is generally adopted today. The use of electricity prices to guide users' electricity consumption behavior has played a certain role in the peak-cutting of the grid. However, due to the fact that the electricity consumption characteristics of residential areas are too remarkable, difference between the peaks and valleys of the power grid is still very large. But the EV charging load has a strong elastic space compared with other loads, thus the electricity price guides the EV owners' charging behavior selection and integrates it into the residential distribution network load, which can better realize the power grid shift peak filling and users and the need for economic operations.

3. Dynamic trading mechanism

3.1. Pricing principle
Utility theory is an important theory in economics. It is a theory exist when the decision makers making decision-making options. Decision-making is mainly influenced by the subjective consciousness of leaders. The charging behavior of electric vehicles is mainly influenced by the

decision of the owner. Different sensitivity levels for charging time and charging cost correspond to different decisions. For the charging decision, you can establish n time period charging utility functions as follows:

$$U(X_1, X_2, \cdots X_n) = X_1^{\rho_1} \cdot X_2^{\rho_2} \cdots X_n^{\rho_n} \tag{1}$$

Where U is the utility function, X_n represents the charging load at different time periods and ρ_n is the elastic coefficient.

The new electricity price setting principle is for electric vehicles, which guides the charging and discharging behavior of electric vehicles through electricity prices and thus reduces the load peak-to-valley difference and maximum load value of the residential area.

3.2. Dynamic price based on TOU price

In Section 1.2, the analysis of the load curve of electric vehicles shows that even under the TOU price, the load of electric vehicles still has obvious periodicity with time. Based on the TOU price, this paper formulates the dynamic electricity price based on the actual load situation [7]:

$$P_f = \frac{P_{f0} \cdot D_f}{D_{f\_ave}} \tag{2}$$

$$P_p = \frac{P_{p0} \cdot D_p}{D_{p\_ave}} \tag{3}$$

$$P_g = \frac{P_{g0} \cdot D_g}{D_{g\_ave}} \tag{4}$$

Among them, P_f , P_p , P_g represents the daily peak, flat and valley electricity price predicted after adjustment, while D_f , D_p , D_g represents the load value of peak, flat and valley after adjustment obtained through the load value forecast at the same period before the day. D_{f_ave} , D_{p_ave} , D_{g_ave} represents the average load value of peak period and valley period respectively. The dynamic electricity price development process is shown in the figure 2 below.

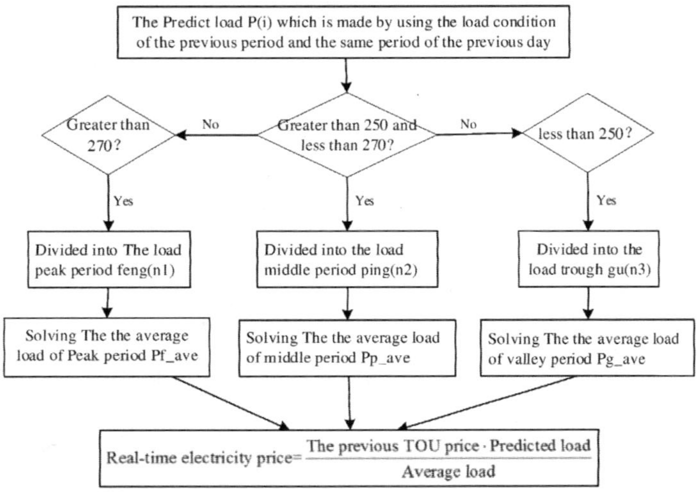

Figure 2. Process for developing dynamic electricity prices

4. User response model based on load elastic demand

4.1. Actual road network mod Load elastic demand matrix

Price elasticity of demand refers to the degree to which the quantity demanded of a commodity reacts to its price change in a certain period of time, representing the change relationship between quantity demanded and price.

$$E = \frac{\partial m}{\partial p} = \frac{p_0}{m_0} \times \frac{dm}{dp} \tag{5}$$

Where E is elasticity of demand, p_0 and p are the electricity price before and after change, m_0 and m respectively correspond to the load under two electricity prices, dp and dm are the difference of the price and the load.

The change in electricity price causes the change of load to be divided into two types: 'self-elastic demand' and 'cross-elastic demand'. 'Self-elastic demand' refers to the sensitivity of a part of the non-transferable load to the price at a fixed time, and the elasticity value is a negative number; 'cross-elastic demand' refers to the load with the transfer ability (the demand for electricity which can be transferred from the peak period to the trough period) has a multi-period sensitivity and its elasticity value is positive.

The elastic matrix E consisting of self-elasticity and cross-elasticity is shown as:

$$E = \begin{bmatrix} \rho_{11} & \rho_{12} & & \rho_{1n} \\ \rho_{21} & \rho_{22} & & \rho_{2n} \\ & & & \\ \rho_{n1} & \rho_{n2} & & \rho_{nn} \end{bmatrix}_{n \times n} \tag{6}$$

4.2. Customer response

Generally speaking, the value of price self-elasticity and price cross-elasticity between time periods in the demand elasticity matrix are within a certain range, or are based on foreign experience. Because the experience only gives a range of the elasticity coefficient, the specific value is quite uncertain, and the user type, the economic situation of each place all have differences, the power demand price elasticity matrix E obtained according to experience cannot be adapted to the users' demand. So we can't take values based on experience. On the contrary, some studies use the linear regression method [8] to perform regression analysis on a certain type of electricity consumption data in a certain region, and then obtain the demand elasticity matrix that satisfies the user's electricity consumption law, and finally obtain the demand response model.

5. Examples

In order to prove the effectiveness of the dynamic electricity price, this section take the load data of a mid-range residential area in a certain city of Anhui Province as the basis, and select an electric vehicle charging station that adopts the TOU price mechanism in the vicinity of the community, the charging station has 30 charging piles with a power of 60 kW. Collect data and select a typical daily load to draw the corresponding load curve, as shown in Figure 3.

A simple analysis of the following parameters is selected for the typical load curve: the maximum peak-to-valley difference in the region before and after the electric vehicle charging station under the TOU tariff is reduced from the original 59.73 MW to 50.59 MW. The variance of load fluctuations was reduced from 335.89 to 269.86. It can be found that the maximum peak-to-valley difference of the load and the variance of the load fluctuation are reduced, but the target effect of the peak-shaving is still not achieved.

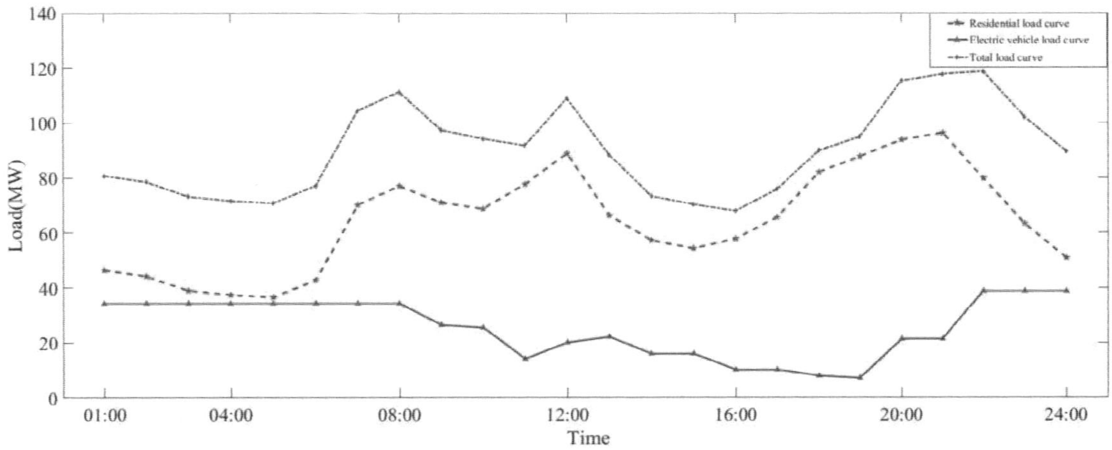

Figure 3. Impact of electric vehicle access on residential area load

5.1. Establish dynamic electricity price and EV load response

Based on the load data of a certain district of Anhui Province and the current TOU electricity price information, a dynamic electricity price mechanism suitable for the electric vehicle charging station near the neighborhood is formulated.

Under the mechanism of the new dynamic electricity price, the peak-to-valley period of the load is no longer a fixed period, but is automatically divided into the corresponding level by comparing the predicted dynamic load value with the threshold of dividing the peak valley, and then according to the second chapter of this paper. Dynamic electricity prices are obtained based on TOU electricity prices. With the forecast of the resident load at 24 hours a day and the original TOU tariff mechanism, the dynamic electricity price is set as shown in the following table 1:

Table 1. Dynamic electricity price

Time	Residential load(MW)	Residential load(MW)	Dynamic electricity price(yuan)	Time	Residential load(MW)	Residential load(MW)	Dynamic electricity price(yuan)
1:00	46.5	0.8769	0.9213	13:00	66.3	1.1048	1.1231
2:00	44.3	0.8769	0.8777	14:00	57.25	1.1048	1.1343
3:00	38.95	0.8769	0.7717	15:00	54.325	1.1048	1.0763
4:00	37.35	0.8769	0.7400	16:00	57.825	1.1048	1.1457
5:00	36.55	0.8769	0.7242	17:00	65.625	1.1048	1.111
6:00	42.85	0.8769	0.8490	18:00	82	1.455	1.4435
7:00	70.175	0.8769	1.2354	19:00	87.775	1.455	1.5452
8:00	76.95	1.1048	1.3547	20:00	93.825	1.455	1.6517
9:00	70.975	1.1048	1.2495	21:00	96.275	1.455	1.6949
10:00	68.75	1.455	1.2103	22:00	79.75	1.1048	1.4035
11:00	77.625	1.455	1.3665	23:00	63.25	1.1048	1.0714
12:00	88.85	1.455	1.5642	24:00	50.85	0.8769	1.0075

Aiming to predict the response behavior of electric vehicles under dynamic electricity price, firstly, using the linear regression method to carry out regression analysis on the load data before and after the implementation of the TOU price, thus can obtain the elastic matrix of the user demand. The matrix can reflect the consumption patterns of users here, and can be used to predict the load response of electric vehicles under the established dynamic electricity price.

The electric vehicle charging station adopted the unified electricity price of 1 yuan/kW·h before adopting the time-of-use electricity price. The TOU price is set as shown in Table 2. The ten-day load data before and after the implementation of the time-of-use electricity price is passed through the price demand matrix. Regression analysis, constructing the demand price elasticity user response model, and obtaining the user load forecast value.

Table 2. Details of TOU price

	Peak load period	Middle load period	Load valley period
Time	09:00-12:00 17:00-22:00	08:00-09:00 12:00-17:00 22:00-23:00	23:00-08:00
Electricity price(yuan/kW·h)	1.455	1.1048	0.8769

5.2. Dynamic electricity price

Under the new dynamic electricity price mechanism, the predicted electric vehicle charging station load is incorporated into the original residential load, and the following load curve is drawn in figure 4. The curve can be used to visually find the fluctuation of the total load curve after implementing the dynamic electricity price. The electricity price has been somewhat flat. In addition, the main load characteristics can be analyzed by data:

1) The maximum peak-to-valley difference is reduced from the original 50.59 MW to 36.87 MW.

2) The maximum load is reduced from the 118.6 MW to 112 MW

3) The variance of load fluctuation is reduced from the original 269.86 to 137.05

4) The total load of electric vehicles increased, but the average user price dropped from the original 1.068 yuan/ kW·h to 1.004 yuan/ kW·h.

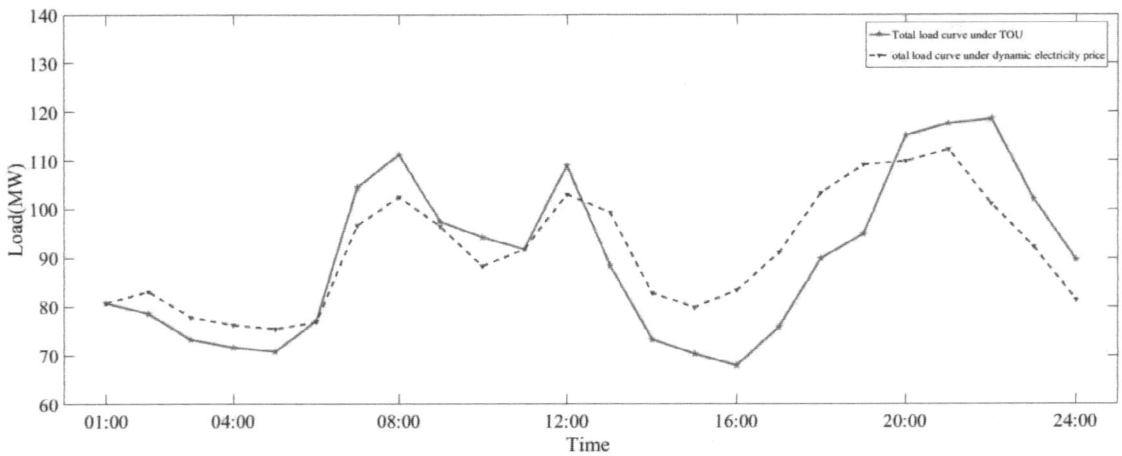

Figure 4. Load curve under different electricity price mechanisms

Through the above model and case analysis, it can bring a new idea to the charging price system of electric vehicles. Taking the electric vehicle charging station open to the outside of residential areas as an example, it can attract other vehicles outside the community by adopting this dynamic electricity price mechanism. Charging during the low load period. The same dynamic mechanism can also be applied to the price guidance of other loads.

6. Conclusions

This paper proposed a dynamic electricity price for electric vehicle charging stations near residential areas. Based on the data-driven, this paper analyzed the difference between the load characteristics of the original residential area before and after the electric vehicle charging load accessing in, and compared the TOU price with the dynamic price. The effectiveness of the dynamic electricity price was verified by the ability of the network to cut the peaks and fill the valleys in the residential area.The most important results of the proposed approach are as follows.

1) The proposed dynamic electricity price is based on the original time-of-use electricity price, and the peak-to-valley period of the load is no longer a fixed period, but is automatically divided to the corresponding level by comparing the predicted dynamic load value with the threshold of dividing the peak valley.

2) The research object is aimed at the electric vehicle charging station near the residential area. Since the electricity usage habits of the residential area are difficult to change, the dynamic electricity price can be moved to attract other non-residential vehicles to charge and then load the peaks and fill the valley of this area.

3) After the guidance of dynamic electricity price, not only the demand for regional peak-cutting and valley filling is better realized, but also the average electricity price of users is reduced, and the economical efficiency of users is improved.

References

[1] Cai H and Chen Q Y 2018 J. Protection and Control of Modern Power Systems. **3** 93-107
[2] Wu C X,Zhang J,Zhang X Y and Chen Z H 2019 J. Power system protection and control. 1-9
[3] Cao Y,Tang S and Li C 2012 J. IEEE Transactions on Smart Grid. **3** 388-393.
[4] Gao Y J,Wang C and Lv M K 2014 J. Electric Power Automation Equipment. **34** 8-12
[5] LIU J,Niu D X,Xing M,Guo L and Zheng S M 2014 J.Power System Technology **38** 1346-51
[6] Chen L and LU S Y 2018 J. Southern Energy Construction **5** 51-58
[7] Liu W and Zhang Z S 2019 J. Proceedings of the CSU-EPSA 1-8
[8] Li C L 2017 D. North China Electric Power University

Research on Inconsistency of Echelon- Use Batteries and Flexible Group Energy Storage Converter

Min Luo [1], Hui Liang [2] and Jinling Meng[1]

[1]Energy Storage Department, Guangdong Power Grid Co. Ltd., Guangzhou 510080, China
[2]School Of Electrical Engineering, Beijing Jiaotong University, Beijing 100044, China

minzi604@139.com

Abstract. In this paper, fifty-five echelon-use batteries were tested and the capacity inconsistency of the batteries are analyzed. The flexible group technology of batteries is studied, and the characteristics and application scenarios of different topologies of energy storage converters are analyzed. Aiming at the problems of high current stress, high loss and high cost in full power control mode, a new energy storage converter topology with integrated primary power control is proposed, and the control method of the topology structure is given. The modeling and Simulation of MATLAB proves that the topology can realize independent control of the charging and discharging current of battery module under highly integrated conditions.

1. Introduction

Because of its energy-saving and environmental protection, electric vehicles have become the main direction of the future development of the automotive industry. With the commercialization of new energy vehicles will be further commercialized, in the next few years a large number of vehicle power batteries will reach their service life and be eliminated, so it is necessary to make rational use of retired electric vehicle batteries.

Retired electric vehicle batteries are commonly echelon-used in energy storage systems for distributed energy access and other occasions [1-2]. However, the consistency of the echelon-use batteries is poor, and the life of the batteries is greatly shortened due to the interaction of traditional batteries. Flexible grouping technology is an effective way to solve the inconsistency of batteries in cascade utilization [3-5]. Flexible group energy storage system can control charge and discharge current independently according to each battery module, which can reduce the requirement of battery capacity consistency and the cost of regrouping, and improve the utilization rate of battery capacity while meeting the requirements of energy storage system.

In order to solve the problem of inconsistent capacity of returned batteries, the capacity tests of 55 batteries with cascade utilization were carried out. The characteristics and application scenarios of the topology structure of cascade battery flexible group energy storage converter are studied. The topology of the original integrated partial power control energy storage converter is proposed, and the control method of the topology is given. The effectiveness of the energy storage converter is verified by MATLAB modeling and simulation.

Content from this work may be used under the terms of the Creative Commons Attribution 3.0 licence. Any further distribution of this work must maintain attribution to the author(s) and the title of the work, journal citation and DOI.
Published under licence by IOP Publishing Ltd

2. Capacity inconsistency analysis of Echelon-use batterries

2.1. Capacity test of echelon-use batteries

In this paper, a certain type of retired electric vehicle batteries from a electric bus has used for more than three years. The nominal voltage/capacity of single battery is 3.2V/20Ah, the upper limit of voltage is 3.65V, the lower limit of voltage is 2V, the specifications of modules are 4 series and 8 parallel, and the nominal voltage/capacity of single module is 12.8V/160Ah. Charging and discharging are carried out at 0.1C. The voltage of any battery module reaches the upper/lower limit of the voltage and stops charging and discharging.

55 battery modules are grouped into 11 groups according to capacity approximation, and the results of capacity testing are shown in Table 1.

Table 1. Summary of capacity test for echelon-use batteries

Group	Module capacity（Ah）				
	Module1	Module2	Module3	Module4	Module5
1	121.560	125.022	126.955	126.957	127.265
2	127.962	128.234	128.475	129.209	129.770
3	131.286	131.460	132.063	133.656	134.874
4	134.919	135.279	135.693	135.919	136.019
5	136.063	136.073	136.090	136.529	136.695
6	136.761	137.237	137.605	137.699	137.900
7	138.265	138.297	138.326	138.712	138.786
8	138.851	139.260	139.460	139.595	139.827
9	139.984	140.030	140.182	140.337	140.528
10	140.690	140.804	140.853	140.992	141.014
11	141.101	141.897	142.006	142.814	144.214

2.2. Analysis of capacity inconsistency between battery modules

There are 55 battery modules in this test. After a long period of use, the battery is aging and capacity declines. The inconsistency of the battery manufacturing process leads to the difference in the initial characteristics of the battery. The difference in the temperature, vibration level and connection between the batteries makes the battery more inconsistent. The inconsistency of battery modules can be obtained by measuring capacity.

The rated capacity of batteries is 160Ah, the average capacity of cascade batteries is 136.146 Ah, and the capacity decline rate is 14.91%. The capacity of batteries after the decline varies greatly. The maximum available capacity of batteries is 144.214Ah, the minimum available capacity of batteries is 121.56Ah and the range of battery capacity is 22.654Ah. The capacity scatter plot, capacity distribution and fitted statistical distribution curves of batteries module for cascade utilization are shown in Figure 1 and 2, respectively. Combining the two graphs, it can be seen that the capacity of battery modules is basically distributed between 125 and 145 Ah, and only one module's capacity falls within the range of 125 Ah. Generally speaking, the distribution of battery module's capacity is relatively centralized. The non-parametric test of the capacity data of 55 battery modules shows that the capacity basically obeys the normal distribution.

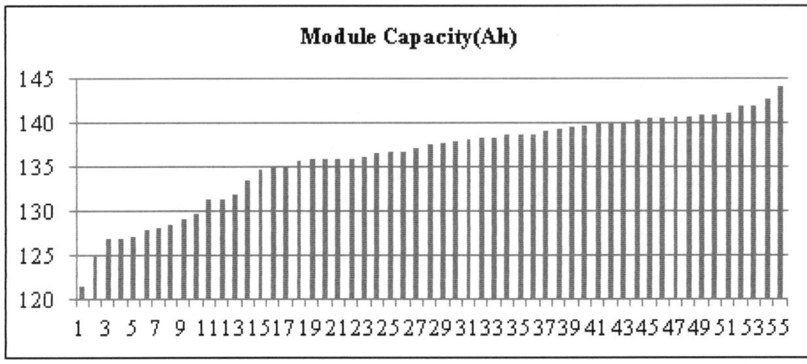

Figure 1. Scatter plot of capacity distribution for 55 battery modules.

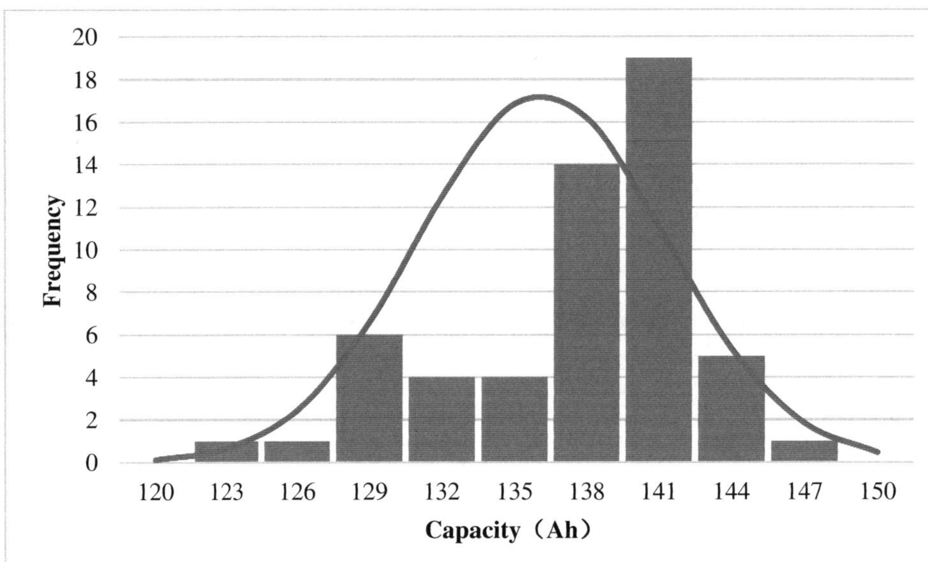

Figure 2. Capacity distribution of 55 battery modules.

3. Research on flexible group energy storage converter

Traditional energy storage system grouping technology is to connect series and parallel batteries to the grid through high-power converters. This grouping method has poor redundancy and obvious short board effect. As can be seen from the above section, after long-term use of batteries, there are obvious differences among batteries. The difference of battery capacity and initial SOC seriously affects the capacity and energy utilization of batteries, and reduces the service life of batteries.

In order to solve this problem, the flexible grouping technology is studied in this paper. Large-scale series-parallel monomer batteries are decomposed into several low-voltage battery modules to form a separate energy storage module. Each module is connected to the grid by flexible group energy storage converter [3-6]. Balanced control is carried out for battery capacity and SOC status within the module, which improves capacity utilization.

3.1. H-bridge cascaded flexible group energy storage converter

Figure 3 is the topology of H-bridge cascaded flexible group energy storage converter. The characteristic of this topology is three-phase AC output structure. Each phase is cascaded by n identical energy storage sub-modules. The DC side of each sub-module is an independent battery module. After the respective H-bridge DC-AC converter, the low voltage AC voltage is generated.

After the modules are connected in series, the high voltage AC voltage required for grid connection is generated. The filter is directly connected to the three-phase AC network.

The topology can independently control the charging and discharging current of the module. Cascade mode can reduce the switching frequency and switching loss. The structure of the system is flexible. It can be directly connected to the AC network to facilitate large-scale centralized energy storage. But its low-voltage application efficiency is low, the cost increases with the total power, the number of modules is large, there are three-phase unbalance problems, and the control is complex. Therefore, H-bridge cascade energy storage system is suitable for high voltage, large capacity, complex battery type, large differential current, high reliability requirements, large-scale centralized energy storage applications.

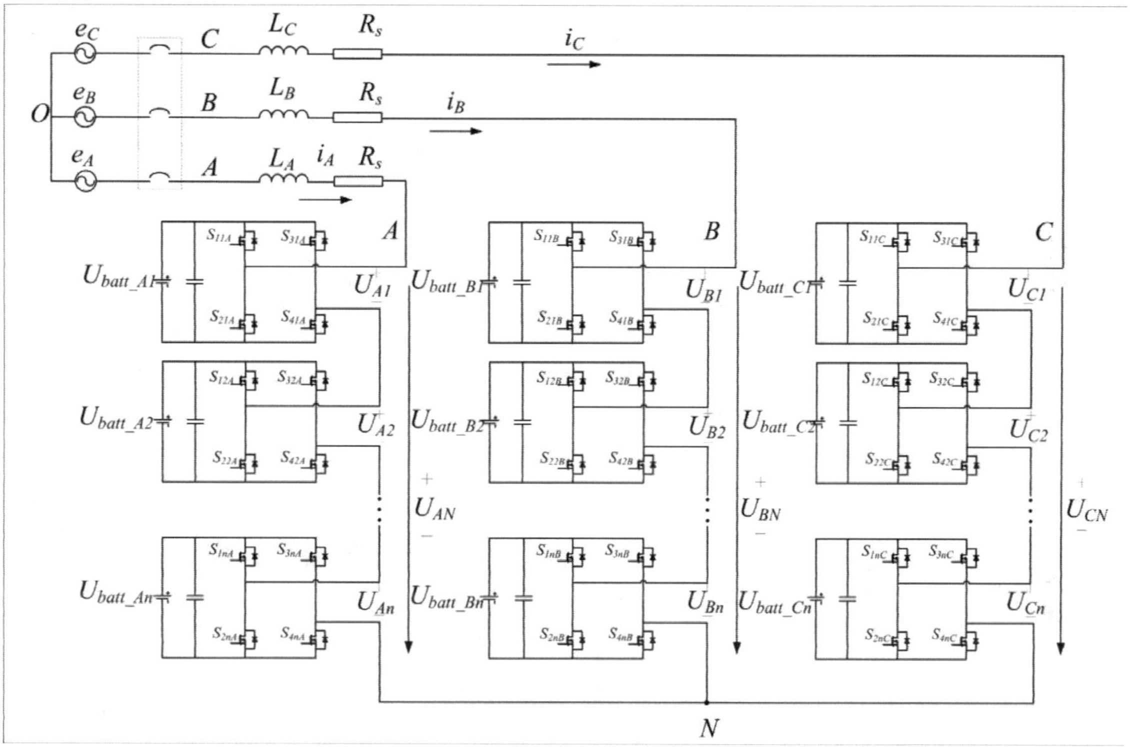

Figure 3. H-bridge cascaded flexible group energy storage converter.

3.2. Modular multilevel flexible group energy storage converter

Figure 4 is the topology of modular multilevel flexible group energy storage converter , which consists of three phase clusters. The two ends of the phase cluster are connected with the common DC bus, and the midpoint of the phase cluster is connected with the AC terminal, which can connect the AC load or the AC network. Each phase cluster consists of two upper and lower bridge arms, inductance of bridge arms and equivalent impedance of line in series. Each bridge arm contains n sub-modules. Each sub-module consists of a half-bridge structure, a supporting capacitor, a bypass switch and a battery module.

This energy storage system can control the current of battery module independently. Its structure can be connected to DC and AC power grids at the same time, realize three-port energy transfer, effectively reduce cost, and have certain fault redundancy function, high reliability. However, its cost and volume increase with the total power. The current of the sub-module contains various frequency components, which makes the power devices lose a lot, and there are circulation problems, and the control strategy is complex. Therefore, modular multilevel battery energy storage system is suitable for high voltage, large capacity, high reliability requirements, AC/DC hybrid microgrid, flexible DC transmission and energy storage applications.

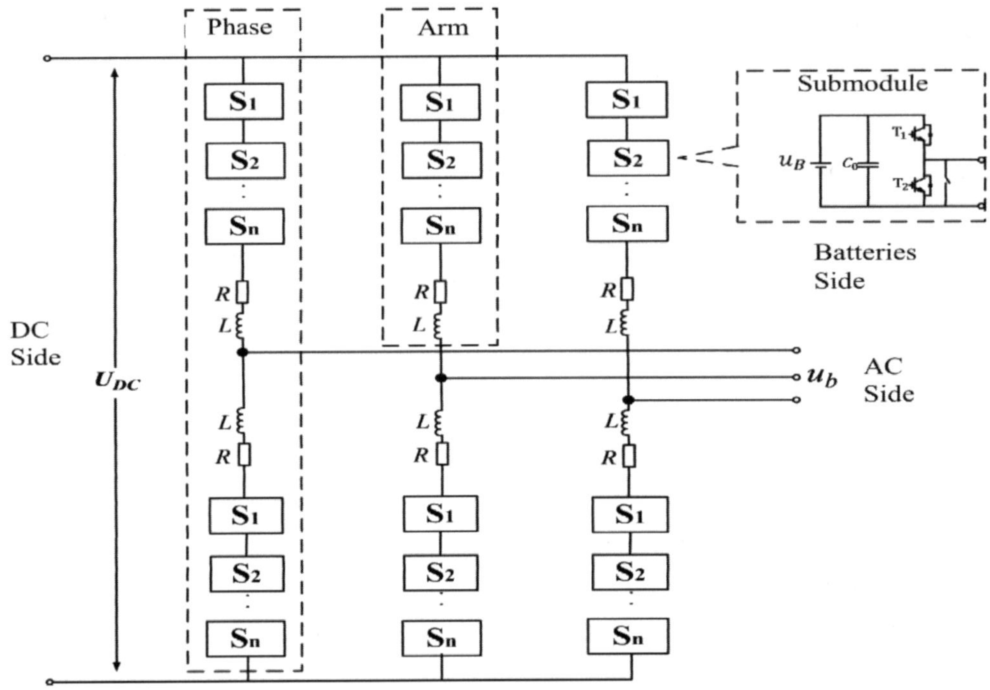

Figure 4. Modular multilevel flexible group energy storage converter.

3.3. DC-DC cascaded flexible group energy storage converter

Figure 5 is the topology of DC-DC cascade flexible group energy storage converter. The left side of the topology is composed of *n* identical DC-DC energy storage modules, which can be connected to the intermediate DC bus or to the AC network through the traditional grid-connected converter. The input side of each energy storage module is an independent battery module. After parallel filter capacitor is connected to DC-DC converter, the controllable DC-DC voltage can be generated. The DC-DC converter can be in the form of Buck-Boost circuit in the figure or other DC-DC converters.

This topology can control module current independently, with fewer switches, simple control and high reliability. Because of the cascade structure, the switch frequency is low, the efficiency is high, and the fault redundancy function can be realized. However, the absolute loss of battery charge and discharge power is large through each conversion module, and its cost and volume increase with the total power. Therefore, this energy storage system is suitable for medium power, medium voltage level, high reliability and redundancy requirements, and complex battery types.

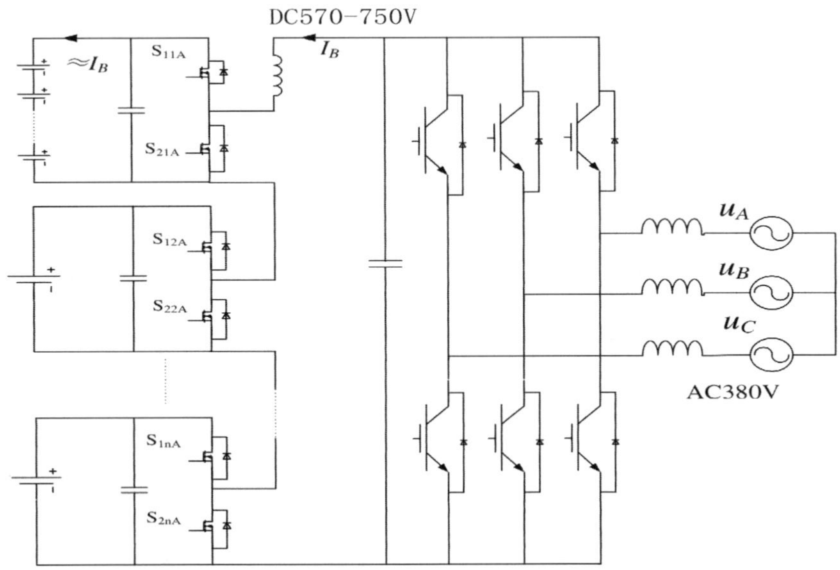

Figure 5. DC-DC cascaded flexible group energy storage converter.

4. Modular independent control energy storage converter with primary integrated

4.1. Topology

The three energy storage systems mentioned above have relatively simple structure and high fault redundancy, which can realize independent control of modules and effectively improve the utilization of battery energy. However, the above three topologies, each module converter for full power control mode, resulting in greater current stress, loss, cost increases accordingly.

From the second section of this paper, it can be seen that there are differences among the battery modules, but within a certain range, it is not necessary to control all the charging and discharging currents independently, only part of the differential current control is needed to achieve the purpose of improving the battery energy utilization. Partial power control has less absolute loss and lower cost than full power control. According to the difference of battery capacity, it can be designed as 10%-20% of the main current. In this paper, a modular independent control energy storage converter with primary integrated is proposed, as shown in Figure 6.

In the middle part of the topology, the batteries are directly connected in series. On the right side, the power control system (PCS) is the main converter. On the left side, the multi-module DC-DC auxiliary converter is the main converter. The battery module interacts with the power grid through PCS, realizes the bidirectional flow of energy between the battery and the grid side, and achieves the functions of "peak shaving and valley filling". The middle part of the energy storage system is composed of cascade batteries connected in series, which can be used as the basis for balanced control of the whole energy storage system through the combination of voltage and current acquisition, SOC estimation and protection control. Left auxiliary converter can be divided into primary side winding converter and multiple secondary side winding converter. One winding is integrated on the primary side and connected to the DC side through the power electronic converter. The secondary side is composed of several sub-conversion modules.

The topology has obvious modularization advantages. The auxiliary converter can be used in many traditional energy storage systems to adjust the total voltage of DC side by changing the number of modules of secondary windings. The total current of the battery is controlled by independent control of the differential current of each module. It not only reduces the cost and loss obviously, but also has low control power and absolute loss, high efficiency, higher utilization rate of battery energy, and

IOP Conf. Series: Materials Science and Engineering **752** (2020) 012024 doi:10.1088/1757-899X/752/1/012024

prolongs the service life of battery. It is suitable for medium and high power and medium voltage applications.

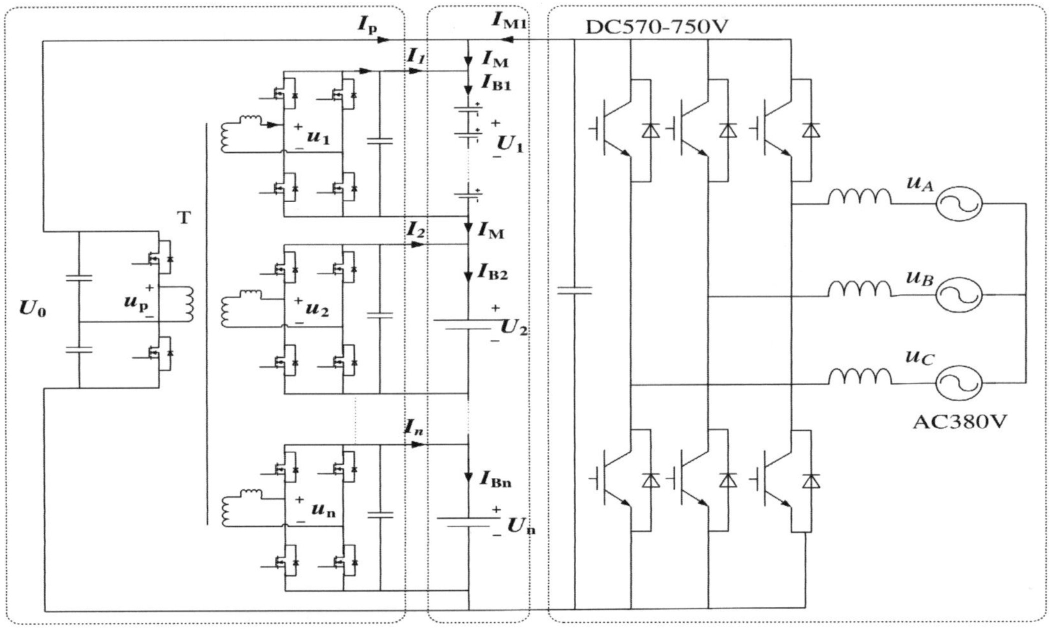

Figure 6. Modular independent control energy storage converter with primary integrated.

4.2. Control method

From the previous analysis, it can be seen that there is a significant difference between the capacity of battery modules and SOC. In order to make each battery module work in the best state, it is necessary to control its current independently. When considering the control strategy of the whole energy storage system, the battery module, PCS and auxiliary converter are integrated. The control block diagram is shown in Figure 7 below.

The control method of the modular energy storage system with primary integrated is as follows:

- PCS charges and discharges series batteries according to grid side demand, and its current value is I_{M1}
- After I_{M1} superimposes the primary current I_p, it acts as the main curcuit current I_M of the series battery pack. After I_M superimposes the differential current I of the secondary modules of the assist converter, the total current I_B of the different battery modules will be the total current I_B of the different battery modules
- Battery Management System (BMS) estimates the state of charge (SOC) of each module by collecting current I_B and voltage U
- The errors of SOC and average SOC of each module are closed-loop controlled respectively, and the output results are taken as the given value of differential current
- After calculating the error between the given value of differential current and the sampled value, the relative phase-shifting angle of each module can be given through current closed-loop control, so as to control the phase-shifting pulse. The relative phase-shifting angle of the pulse determines the magnitude and positive or negative of the transmission power and differential current of the module, thus forming a double closed-loop control

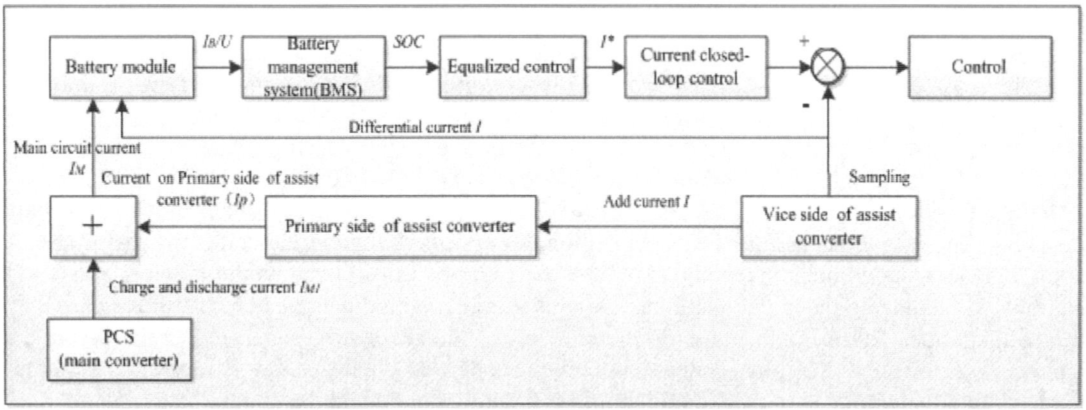

Figure 7. Control logic block diagram.

In summary, after considering the capacity difference between each battery module and SOC, the energy storage system realizes the independent control of the total current of the battery module by controlling the 10%-20% differential current in the main current of the battery. After the positive and negative power of the secondary module is offset, the original power can be neglected. The total control power of the auxiliary converter is small, the absolute loss is small, most of the power is borne by PCS, and the efficiency of the whole machine is low. It is high and easy to control. It can realize grid-connected AC side.

4.3. Simulation verification

This project is based on the modular energy storage system with primary integrated. Matlab/Simulink is used to verify the principle. As shown in Figure 8,When the module 1 changes from the discharge mode to the charging mode, the original side current, the average current and the differential current are shown in Figure 8 below. The new steady state value of the original winding and module 1 current can be obtained, and the remaining modules remain independent control. The independent control of the topology is proved.At the same time, compared with the traditional full power modular control energy storage system, the control current of the system is small and the loss is small.

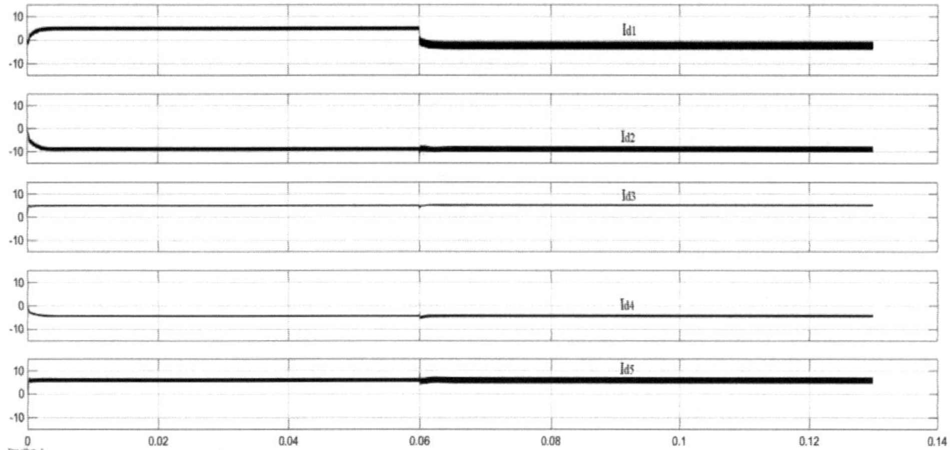

Figure 8. Current control chart of module 1-5 in the modular independent control energy storage converter with primary integrated.

5. Conclusion

By controlling each power converter module, the flexible group energy storage converter described in this paper realizes the independent control of the charging and discharging current of the battery module and improves the usable energy of the battery pack. The first three topologies belong to the full-power independent control system. They are relatively simple, high fault redundancy, high reliability and large independent regulation range. However, the loss, cost and volume of the converter also increase. They are suitable for applications with high voltage, large capacity and large battery difference. The modular independent control energy storage converter with primary integrated proposed in this paper belongs to a part of power independent control system. It realizes module separation and controls the differential current of the battery module independently. It has low cost and small volume. The original power is the sum of all the secondary edges. After the positive and negative power offset, the original power can be neglected, and the efficiency is high. It is suitable for the battery energy storage system of medium and high power and medium voltage level.

Acknowledgements

This work was financially supported by the Guangdong Power Grid Co., Ltd. (Grant NO. GDKJXM20161883).

References

[1] DEBNATH U K, AHMAD I, HABIBI D. Quantifying economic benefits of second life batteries of gridable vehicles in the smart grid [J]. International Journal of Electrical Power & Energy Systems, 2014, 63: 577-587.

[2] Viswanathan, Vilayanur V, Michael K M. Second use of transportation batteries: maximizing the value of batteries for transportation and grid services[J]. IEEE Transactions on Vehicular Technology, 2011, 60(7): 2963-2970.

[3] SERBAN I, MARINESCU C. A look at the role and main topologies of battery energy storage systems for integration in autonomous microgrids[C] // 12th International Conference on Optimization of Electrical and Electronic Equipment (OPTIM), May 20-22, 2010, Basov: 9p.

[4] Hafez A A A. Seven-level three phase cascaded H-bridge inverter with a single DC source[J]. ARPN Journal of Engineering and Applied Sciences, 2012, 7(12): 1546-1554.

[5] Laxman Maharjan, Tsukasa Yamagishi, Hirofumi Akagi. Active-power control of individual converter cells for a battery energy storage system based on a multilevel cascade PWM concenter[J]. IEEE Transactions on Power Electronics, 2012, 27(3):1099-1107.

[6] LIN Y S, HU K W, YEH T H, et al. An electric-vehicle IPMSM drive with interleaved front-end DC/DC converter[J]. IEEE Transactions on Vehicular Technology, 2016, 65(6): 4493-4504.

AUTHOR INDEX